SpringerBriefs in Mathematics

SpringerBriefs present concise summaries of cutting-edge research and practical applications across a wide spectrum of fields. Featuring compact volumes of 50 to 125 pages, the series covers a range of content from professional to academic. Briefs are characterized by fast, global electronic dissemination, standard publishing contracts, standardized manuscript preparation and formatting guidelines, and expedited production schedules.

Typical topics might include:

- A timely report of state-of-the art techniques
- A bridge between new research results, as published in journal articles, and a contextual literature review
- A snapshot of a hot or emerging topic
- An in-depth case study
- A presentation of core concepts that students must understand in order to make independent contributions

SpringerBriefs in Mathematics showcases expositions in all areas of mathematics and applied mathematics. Manuscripts presenting new results or a single new result in a classical field, new field, or an emerging topic, applications, or bridges between new results and already published works, are encouraged. The series is intended for mathematicians and applied mathematicians. All works are peer-reviewed to meet the highest standards of scientific literature.

Titles from this series are indexed by Scopus, Web of Science, Mathematical Reviews, and zbMATH.

Anton Petrunin

Pure Metric Geometry

Anton Petrunin
Department of Mathematics
Pennsylvania State University
University Park, PA, USA

ISSN 2191-8198 ISSN 2191-8201 (electronic)
SpringerBriefs in Mathematics
ISBN 978-3-031-39161-3 ISBN 978-3-031-39162-0 (eBook)
https://doi.org/10.1007/978-3-031-39162-0

This Springer imprint is published by the registered company Springer Nature Switzerland AG
The registered company address is: Gewerbestrasse 11, 6330 Cham, Switzerland

Preface

This text can serve as an introductory part for a variety of courses in metric geometry. Here is a graph of essential dependencies of the lectures; some statements (mostly exercises) add more dependencies, but they can be ignored. The necessary definitions are introduced in Lecture 1. In Lecture 2, we discuss the Urysohn space. In Lecture 3, we discuss injective spaces. In Lecture 4, we introduce the Hausdorff metric. In Lectures 5 and 6, we discuss two types of convergences of metric spaces—the Gromov–Hausdorff limit and ultralimit.

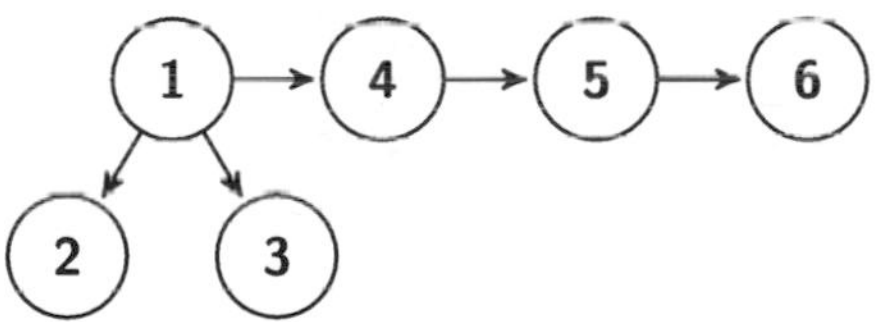

Applications are given only as illustrations. We stick to domestic affairs of metric spaces, keeping away from any extra structure. (Adding an extra structure brings an extra tool and often opens a huge field for development. The examples include Alexandrov geometry, geometric group theory, metric-measure spaces, and optimal transport.)

The only prerequisite is interest in the subject, but any knowledge of classical geometry, differential geometry, topology, and real analysis will be useful.

These notes are based on the minicourse given at SPbSU (Fall 2022) and the introductory part of a course at PSU (Spring 2020). The latter included additional material from [1, 47, 56]. A part of the text is a compilation from [1, 2, 50, 54, 55] and its drafts.

I want to thank Sergei Ivanov, Urs Lang, Alexander Lytchak, Rostislav Matveyev, Julien Melleray, and Sergio Zamora Barrera for their help. The present work is partially supported by NSF grant DMS-2005279, the Simons Foundation grant #584781, and Minobrnauki of Russia, grant #075-15-2022-289.

University Park, PA, USA Anton Petrunin

Contents

Lecture 1
Definitions

In this lecture, we remind several definitions related to metric spaces and fix some conventions.

This lecture is self-contained, but it is written for students with some prior knowledge of metric spaces; an introduction to general topology is sufficient but not necessary. For a more detailed introduction, we recommend the first couple of chapters in the book by Dmitri Burago, Yuri Burago, and Sergei Ivanov [13].

A Metric spaces

The distance between two points x and y in a metric space $\mathcal{X}$ will be denoted by $|x-y|$ or $|x-y|_{\mathcal{X}}$. The latter notation is used if we need to emphasize that the distance is taken in the space $\mathcal{X}$.

Let us recall the definition of metric.

1.1. Definition. A *metric* on a set $\mathcal{X}$ is a real-valued function $(x,y)\mapsto|x-y|_{\mathcal{X}}$ that satisfies the following conditions for any $x,y,z\in\mathcal{X}$:

(a) $|x-y|_{\mathcal{X}}\geqslant 0$.
(b) $|x-y|_{\mathcal{X}}=0 \iff x=y$.
(c) $|x-y|_{\mathcal{X}}=|y-x|_{\mathcal{X}}$.
(d) $|x-y|_{\mathcal{X}}+|y-z|_{\mathcal{X}}\geqslant|x-z|_{\mathcal{X}}$.

Recall that a *metric space* is a set with a metric on it. The elements of the set are called *points*. Most of the time we keep the same notation for the metric space and its underlying set; the latter can be denoted by $\underline{\mathcal{X}}$ if needed.

A. Petrunin, *Pure Metric Geometry*, SpringerBriefs in Mathematics,
https://doi.org/10.1007/978-3-031-39162-0_1

Given radius $R \in [0, \infty]$ and center $x \in \mathcal{X}$, the sets

$$\mathrm{B}(x, R) = \{\, y \in \mathcal{X} \,:\, |x - y| < R \,\},$$
$$\overline{\mathrm{B}}[x, R] = \{\, y \in \mathcal{X} \,:\, |x - y| \leqslant R \,\}$$

are called, respectively, the open and the closed balls. The notations $\mathrm{B}(x, R)_{\mathcal{X}}$ and $\overline{\mathrm{B}}[x, R]_{\mathcal{X}}$ might be used if we need to emphasize that these balls are taken in the metric space $\mathcal{X}$.

1.2. Exercise. Show that the following inequality

$$|p - q|_{\mathcal{X}} + |x - y|_{\mathcal{X}} \leqslant |p - x|_{\mathcal{X}} + |p - y|_{\mathcal{X}} + |q - x|_{\mathcal{X}} + |q - y|_{\mathcal{X}}$$

holds for any four points p, q, x, and y in a metric space $\mathcal{X}$.

B Topology

The standard calculus definitions of closed and open sets, continuous functions, and converging sequences admit straightforward generalizations in the context of metric spaces.

1.3. Exercise. Let x be a point in a metric space $\mathcal{X}$. Show that the distance function $\mathrm{dist}_x : \mathcal{X} \to \mathbb{R}$ defined by

$$\mathrm{dist}_x : y \mapsto |x - y|$$

is continuous.

1.4. Exercise. Let A and B be two disjoint closed sets in a metric space $\mathcal{X}$. Construct a continuous function $f : \mathcal{X} \to [0, 1]$ such that $A = f^{-1}\{0\}$ and $B = f^{-1}\{1\}$.

1.5. Advanced exercise. Let $f : A \to \mathbb{R}$ be a continuous function defined on a closed set A in a metric space $\mathcal{X}$. Show that it admits a continuous extension to the whole space; that is, there is a continuous function $F : \mathcal{X} \to \mathbb{R}$ such that $F(a) = f(a)$ for any $a \in A$.

C Variations

Pseudometris A metric for which the distance between two distinct points can be zero is called a semimetric (also known as pseudometric). In other words, to define semimetric, we need to remove condition (b) from 1.1.

Assume $\mathcal{X}$ is a semimetric space. Consider an equivalence relation $\sim$ on $\mathcal{X}$ defined by

$$x \sim y \qquad \Longleftrightarrow \qquad |x - y| = 0.$$

Note that if $x \sim x'$, then $|y - x| = |y - x'|$ for any $y \in \mathcal{X}$. Thus, $|* - *|$ defines a metric on the quotient $\mathcal{X}/{\sim}$. The so-obtained metric space, say $\mathcal{X}'$, is called the *corresponding metric space* for the semimetric space $\mathcal{X}$.

This construction shows that nearly any question about semimetric spaces can be reduced to a question about genuine metric spaces. Often we do not distinguish between a semimetric space $\mathcal{X}$ and its corresponding metric space $\mathcal{X}'$.

∞-Metrics One may also consider metrics with values in $[0, \infty]$; that is, we allow infinite distance between points. We might call them *∞-metrics*, but most of the time we use the term *metric*.

The following construction shows how to reduce questions about ∞-metrics to genuine metrics.

Let

$$x \overset{\infty}{\sim} y \qquad \Longleftrightarrow \qquad |x - y| < \infty;$$

it defines another equivalence relation on $\mathcal{X}$. The equivalence class of a point $x \in \mathcal{X}$ will be called the *metric component* of x; it will be denoted by $\mathcal{X}_x$. Note that

$$\mathcal{X}_x = \mathrm{B}(x, \infty)_{\mathcal{X}};$$

that is, the metric component of x is the open ball centered at x and radius ∞.

If $\{\mathcal{X}_\alpha\}$ is a collection of metric spaces, then *disjoint union* $\boldsymbol{X} = \bigsqcup_\alpha \mathcal{X}_\alpha$ will be considered with a natural metric defined by

$$|x - y|_{\boldsymbol{X}} := \begin{cases} |x - y|_{\mathcal{X}_\alpha} & \text{if} \quad x, y \in \mathcal{X}_\alpha \quad \text{for some} \quad \alpha, \\ \infty & \text{otherwise.} \end{cases}$$

It follows that any ∞-metric space is a disjoint union of genuine metric spaces—the metric components of the original ∞-metric space.

1.6. Exercise. Given two sets A and B on the plane, set

$$|A - B| = \mu(A \triangle B),$$

where μ denotes the Lebesgue measure and $\triangle$ denotes symmetric difference

$$A \triangle B := (A \cup B) \setminus (B \cap A) = (A \setminus B) \cup (B \setminus A).$$

(a) Show that $|{*}-{*}|$ is a semimetric on the set of bounded closed subsets.
(b) Show that $|{*}-{*}|$ is an ∞-metric on the set of all open subsets.

D Maximal metric and gluing

Maximal metric Let $\{\,|-|_\alpha\}$ be a family of ∞-semimetrics on a fixed set. Observe that

$$|x-y| := \sup_\alpha\{\,|x-y|_\alpha\,\}$$

defines an ∞-semimetric; it is called the maximal metric of the family.

Gluing Suppose $\sim$ is an equivalence relation on an ∞-semimetric space $\mathcal{X}$. Given $x\in\mathcal{X}$, denote by $[x]$ its equivalence class in the quotient $\mathcal{X}/\sim$. Consider all ∞-semimetrics $|-|_\alpha$ on $\mathcal{X}/\sim$ such that the maps $\mathcal{X}\to\mathcal{X}/\sim$ defined by $x\mapsto[x]$ is short; that is,

$$|[x]-[x']|_\alpha \leqslant |x-x'|_{\mathcal{X}}$$

for any $x,x'\in\mathcal{X}$. Let us equip $\mathcal{X}/\sim$ with the maximal metric of this family; in general, it is an ∞-semimetric. The space $\mathcal{Z}$ that corresponds to the obtained ∞-semimetric space is called gluing of $\mathcal{X}$ along $\sim$.

This definition can be applied to a disjoint union of spaces; this way we can glue an arbitrary collection of metric spaces.

Note that any partially defined map φ from $\mathcal{X}$ to $\mathcal{Y}$ defines a minimal equivalence relation on $\mathcal{X}\sqcup\mathcal{Y}$ such that $x\sim\varphi(x)$; the corresponding gluing space is called gluing along φ.

The following exercise shows that metric gluing and the corresponding topological gluing might have different topologies.

1.7. Exercise. Construct a homeomorphism $\varphi\colon[0,1]\to[0,1]$ such that gluing of two unit intervals $[0,1]$ along φ is a one-point metric space.

E Completeness

A metric space $\mathcal{X}$ is called complete if every Cauchy sequence of points in $\mathcal{X}$ converges in $\mathcal{X}$.

1.8. Exercise. Suppose that ρ is a positive continuous function on a complete metric space $\mathcal{X}$ and $\varepsilon > 0$. Show that there is a point $x \in \mathcal{X}$ such that

$$\rho(x) < (1+\varepsilon)\cdot\rho(y)$$

for any point $y \in \mathrm{B}(x, \rho(x))$.

Most of the time we will assume that a metric space is complete. The following construction produces a complete metric space $\bar{\mathcal{X}}$ for any given metric space $\mathcal{X}$.

Completion Given a metric space $\mathcal{X}$, consider the set $\mathcal{C}$ of all Cauchy sequences in $\mathcal{X}$. Note that for any two Cauchy sequences (x_n) and (y_n) the right-hand side in ➀ is defined; moreover, it defines a semimetric on $\mathcal{C}$

➀ $$|(x_n) - (y_n)|_{\mathcal{C}} := \lim_{n\to\infty} |x_n - y_n|_{\mathcal{X}}.$$

The corresponding metric space is called the completion of $\mathcal{X}$; it will be denoted by $\bar{\mathcal{X}}$.

For each point $x \in \mathcal{X}$, one can consider a constant sequence $x_n = x$ which is Cauchy. It defines a natural inclusion map $\mathcal{X} \hookrightarrow \bar{\mathcal{X}}$. It is easy to check that this map is distance-preserving. In particular, we can (and will) consider $\mathcal{X}$ as a subset of $\bar{\mathcal{X}}$.

Note that $\mathcal{X}$ is a dense subset in its completion $\bar{\mathcal{X}}$.

1.9. Exercise. Show that the completion of a metric space is complete.

F G-delta sets

1.10. Baire's theorem. *For any sequence $\Omega_1, \Omega_2, \dots$ of open dense subsets in a complete metric space, the intersection $\bigcap_{n\in\mathbb{N}} \Omega_n$ is dense.*

A subset is called a G-delta if it can be presented as an intersection of a countable number of open subsets. Note that by Baire's theorem, a countable intersection of dense G-delta sets is a dense G-delta set—in particular, it is nonempty. Therefore, we are allowed to say that a dense G-delta set contains *most* of the points in a complete metric space.

Proof. We may assume that the space is nonempty; otherwise, there is nothing to prove.

Given a closed ball $\overline{\mathrm{B}}[p_0, R_0]$, let us apply induction to construct a nested sequence of closed balls

$$\overline{\mathrm{B}}[p_0, R_0] \supset \overline{\mathrm{B}}[p_1, R_1] \supset \overline{\mathrm{B}}[p_2, R_2] \supset \dots$$

such that $\overline{\mathrm{B}}[p_n, R_n] \subset \Omega_n$ and $R_n > 0$ for each $n \geqslant 1$. Assume $\overline{\mathrm{B}}[p_{n-1}, R_{n-1}]$ is already constructed. Since Ω_n is dense, we can choose a closed ball $\overline{\mathrm{B}}[p_n, R_n] \subset \Omega_n \cap \overline{\mathrm{B}}[p_{n-1}, R_{n-1}]$.

Note that we can assume that $R_n < \frac{1}{n}$ for each $n \geqslant 1$. In this case, the sequence $p_1, p_2, \ldots$ is Cauchy; therefore, it is converging. Observe that its limit p_∞ belongs to each Ω_n. It follows that any closed ball $\overline{\mathrm{B}}[p_0, R_0]$ contains a point in $\bigcap_{n \in \mathbb{N}} \Omega_n$, hence the result. □

G Compact spaces

Let us recall a few statements about compact metric spaces.

1.11. Definition. A metric space $\mathcal{K}$ is compact if and only if one of the following equivalent conditions holds:

(a) Every open cover of $\mathcal{K}$ has a finite subcover.
(b) Every sequence of points in $\mathcal{K}$ has a subsequence that converges in $\mathcal{K}$.
(c) The space $\mathcal{K}$ is complete and totally bounded; that is, for any $\varepsilon > 0$, the space $\mathcal{K}$ admits a finite cover by open ε-balls.

Totally bounded spaces are also called precompact. Note that a *space is precompact if and only if its completion is compact.*

1.12. Lebesgue lemma. *Let $\mathcal{K}$ be a compact metric space. Then for any open cover of $\mathcal{K}$, there is $\varepsilon > 0$ such that any ε-ball in $\mathcal{K}$ lies in an element of the cover.*

The value ε is called a Lebesgue number of the covering.

A subset N of a metric space $\mathcal{K}$ is called ε-net if any point $x \in \mathcal{K}$ lies at the distance less than ε from a point in N. More generally, a subset N is called an ε-net of a subset $S \subset \mathcal{K}$ if any point $x \in S$ lies at the distance less than ε from a point in N.

Note that totally bounded spaces can be defined as spaces that admit a finite ε-net for any $\varepsilon > 0$.

1.13. Exercise. Show that a space $\mathcal{K}$ is totally bounded if and only if it contains a compact ε-net for any $\varepsilon > 0$.

Let $\operatorname{pack}_\varepsilon \mathcal{X}$ be the exact upper bound on the number of points $x_1, \ldots, x_n \in \mathcal{X}$ such that $|x_i - x_j| \geqslant \varepsilon$ if $i \neq j$.

If $n = \operatorname{pack}_\varepsilon \mathcal{X} < \infty$, then the collection of points $x_1, \ldots, x_n$ is called a maximal ε-packing. If $\mathcal{X}$ is a length space (see Sect. L), then n is the maximal number of disjoint open $\frac{\varepsilon}{2}$-balls in $\mathcal{X}$.

1.14. Exercise. Show that any maximal ε-packing is an ε-net. Conclude that a complete space $\mathcal{X}$ is compact if and only if $\operatorname{pack}_\varepsilon \mathcal{X} < \infty$ for any $\varepsilon > 0$.

1.15. Exercise. Let $\mathcal{K}$ be a compact metric space and $f : \mathcal{K} \to \mathcal{K}$ be a distance-noncontracting map. Prove that f is an isometry; that is, f is a distance-preserving bijection.

A metric space $\mathcal{X}$ is called locally compact if any point in $\mathcal{X}$ admits a compact neighborhood; equivalently, for any point $x \in \mathcal{X}$, a closed ball $\overline{\mathrm{B}}[x, r]$ is compact for some $r > 0$.

H Proper spaces

A metric space $\mathcal{X}$ is called proper if all closed bounded sets in $\mathcal{X}$ are compact. Note that $\mathcal{X}$ is proper if for some (and therefore any) point $p \in \mathcal{X}$ and any $R < \infty$, the closed ball $\overline{\mathrm{B}}[p, R]_{\mathcal{X}}$ is compact.

Recall that a function $f : \mathcal{X} \to \mathbb{R}$ is proper if, for any compact set $K \subset \mathbb{R}$, its inverse image $f^{-1}(K)$ is compact. Observe that $\mathcal{X}$ is proper if and only if the function $\mathrm{dist}_p : \mathcal{X} \to \mathbb{R}$ is proper for some (and therefore any) point $p \in \mathcal{X}$.

1.16. Exercise. Give an example of a metric space that is locally compact but not proper.

I Geodesics

Let $\mathcal{X}$ be a metric space and $\mathbb{I}$ a real interval. A distance-preserving map $\gamma : \mathbb{I} \to \mathcal{X}$ is called a geodesic;[1] in other words, $\gamma : \mathbb{I} \to \mathcal{X}$ is a geodesic if

$$|\gamma(s) - \gamma(t)|_{\mathcal{X}} = |s - t|$$

for any pair $s, t \in \mathbb{I}$.

If $\gamma : [a, b] \to \mathcal{X}$ is a geodesic such that $p = \gamma(a)$, $q = \gamma(b)$, then we say that γ is a geodesic from p to q. In this case, the image of γ is denoted by $[pq]$, and, with abuse of notations, we also call it a geodesic. We may write $[pq]_{\mathcal{X}}$ to emphasize that the geodesic $[pq]$ is in the space $\mathcal{X}$.

In general, a geodesic from p to q need not exist, and if it exists, it need not be unique. However, once we write $[pq]$, we assume that we have chosen such geodesic.

A geodesic path is a geodesic with constant-speed parametrization by the unit interval $[0, 1]$.

A metric space is called geodesic if any pair of its points can be joined by a geodesic.

An ∞-metric space $\mathcal{X}$ is called geodesic if each metric component of $\mathcal{X}$ is geodesic.

---

[1] Others call it differently: *shortest path*, *minimizing geodesic*. Also, note that the meaning of the term *geodesic* is different from what is used in Riemannian geometry, altho they are closely related.

1.17. Exercise. Let f be a centrally symmetric positive continuous function on $\mathbb{S}^2$. Given two points $x, y \in \mathbb{S}^2$, set

$$\|x - y\| = \int_{B(x,\frac{\pi}{2})\setminus B(y,\frac{\pi}{2})} f.$$

Show that $(\mathbb{S}^2, \|* - *\|)$ is a geodesic space, and the geodesics in $(\mathbb{S}^2, \|* - *\|)$ run along great circles of $\mathbb{S}^2$.

J Metric trees

A geodesic space $\mathcal{T}$ is called a *metric tree* if any two points in $\mathcal{T}$ are connected by a unique geodesic, and the union of any two geodesics $[xy]_{\mathcal{T}}$, and $[yz]_{\mathcal{T}}$ contains the geodesic $[xz]_{\mathcal{T}}$.

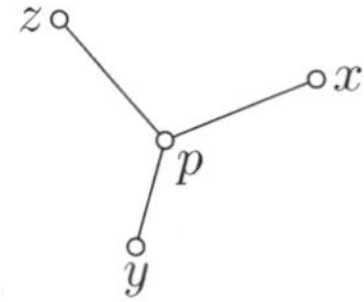

The latter means that any triangle in $\mathcal{T}$ is a *tripod*. That is, for any three points x, y, and z, there is a point p such that

$$[xy] \cup [yz] \cup [zx] = [px] \cup [py] \cup [pz].$$

1.18. Exercise. Let p, x, y, and z be points in a metric tree:

(a) Consider three numbers

$$\begin{aligned} a &= |p - x| + |y - z|, \\ b &= |p - y| + |z - x|, \\ c &= |p - z| + |x - y|. \end{aligned}$$

Suppose that $a \leqslant b \leqslant c$. Show that $b = c$.

(b) Consider three numbers

$$\begin{aligned} \alpha &= \tfrac{1}{2} \cdot (|p - y| + |p - z| - |y - z|), \\ \beta &= \tfrac{1}{2} \cdot (|p - x| + |p - z| - |x - z|), \\ \gamma &= \tfrac{1}{2} \cdot (|p - x| + |p - y| - |x - y|). \end{aligned}$$

Suppose that $\alpha \leqslant \beta \leqslant \gamma$. Show that $\alpha = \beta$.

The set

$$S(p, r)_{\mathcal{X}} = \{ x \in \mathcal{X} : |p - x|_{\mathcal{X}} = r \}$$

will be called a sphere with center p and radius r in a metric space $\mathcal{X}$.

1.19. Exercise. Show that spheres in metric trees are ultrametric spaces. That is,

$$|x - z| \leqslant \max\{ |x - y|, |y - z| \}$$

for any $x, y, z \in S(p, r)_{\mathcal{T}}$.

K Length

A curve is defined as a continuous map from a real interval $\mathbb{I}$ to a metric space. If $\mathbb{I} = [0, 1]$, then the curve is called a path.

1.20. Definition. Let $\mathcal{X}$ be a metric space and $\alpha \colon \mathbb{I} \to \mathcal{X}$ be a curve. We define the length of α as

$$\operatorname{length} \alpha := \sup_{t_0 \leqslant t_1 \leqslant \ldots \leqslant t_n} \sum_i |\alpha(t_i) - \alpha(t_{i-1})|.$$

A curve α is called rectifiable if $\operatorname{length} \alpha < \infty$.

1.21. Theorem. *Length is a lower semi-continuous with respect to the pointwise convergence of curves.*

More precisely, assume that a sequence of curves $\gamma_n \colon \mathbb{I} \to \mathcal{X}$ in a metric space $\mathcal{X}$ converges pointwise to a curve $\gamma_\infty \colon \mathbb{I} \to \mathcal{X}$; that is, for any fixed $t \in \mathbb{I}$, we have $\gamma_n(t) \to \gamma_\infty(t)$ as $n \to \infty$. Then

② $$\varliminf_{n \to \infty} \operatorname{length} \gamma_n \geqslant \operatorname{length} \gamma_\infty.$$

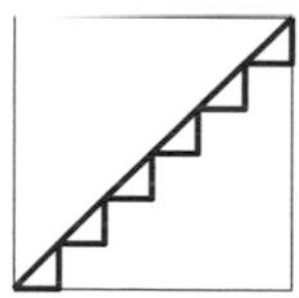

Note that the inequality ② might be strict. For example, the diagonal γ_∞ of the unit square can be approximated by stairs-like polygonal curves γ_n with sides parallel to the sides of the square (γ_6 is on the picture). In this case,

$$\operatorname{length}\gamma_\infty = \sqrt{2} \quad \text{and} \quad \operatorname{length}\gamma_n = 2$$

for any n.

Proof. Fix a sequence $t_0 \leqslant \cdots \leqslant t_k$ in $\mathbb{I}$. Set

$$\Sigma_n := |\gamma_n(t_0) - \gamma_n(t_1)| + \cdots + |\gamma_n(t_{k-1}) - \gamma_n(t_k)|.$$
$$\Sigma_\infty := |\gamma_\infty(t_0) - \gamma_\infty(t_1)| + \cdots + |\gamma_\infty(t_{k-1}) - \gamma_\infty(t_k)|.$$

Note that for each i we have

$$|\gamma_n(t_{i-1}) - \gamma_n(t_i)| \to |\gamma_\infty(t_{i-1}) - \gamma_\infty(t_i)|,$$

and therefore,

$$\Sigma_n \to \Sigma_\infty$$

as $n \to \infty$. Note that

$$\Sigma_n \leqslant \operatorname{length}\gamma_n$$

for each n. Hence,

$$\varliminf_{n\to\infty} \operatorname{length}\gamma_n \geqslant \Sigma_\infty.$$

Since the partition was arbitrary, applying the definition of length, we get ②. □

1.22. Exercise. Show that most of 1-Lipschitz paths in the plane have length 1.

More precisely, consider the space $\mathcal{P}$ of 1-Lipschitz paths in the plane; that is, all paths $a\colon [0,1] \to \mathbb{R}^2$ such that $|a(t_0) - a(t_1)| \leqslant |t_0 - t_1|$ for any t_0 and t_1. Equip $\mathcal{P}$ with the metric defined by

$$|a - b| := \sup\left\{\, |a(t) - b(t)| \;:\; t \in [0,1] \,\right\}.$$

Show that a dense G-delta set of paths in $\mathcal{P}$ has length 1.

L Length spaces

Let $\mathcal{X}$ be a metric space. If for any $\varepsilon > 0$ and any pair of points $x, y \in \mathcal{X}$, there is a path α connecting x to y such that

$$\operatorname{length}\alpha < |x - y| + \varepsilon,$$

then $\mathcal{X}$ is called a length space and the metric on $\mathcal{X}$ is called a length metric.

An ∞-metric space is a length space if each of its metric components is a length space. In other words, if $\mathcal{X}$ is an ∞-metric space, then in the above definition we assume in addition that $|x - y|_{\mathcal{X}} < \infty$.

Note that any geodesic space is a length space. The following example shows that the converse does not hold.

1.23. Example. Set $\mathbb{I}_n = [0, 1 + \frac{1}{n}]$ for every natural n. Suppose a space $\mathcal{X}$ is obtained by gluing intervals $\{\mathbb{I}_n\}$, where the left ends are glued to p and the right ends to q.

Observe that the space $\mathcal{X}$ carries a natural complete length metric with respect to which $|p - q|_{\mathcal{X}} = 1$, but there is no geodesic connecting p to q.

1.24. Exercise. Give an example of a complete length space $\mathcal{X}$ such that no pair of distinct points in $\mathcal{X}$ can be joined by a geodesic.

Directly from the definition, it follows that if $\alpha\colon [0, 1] \to \mathcal{X}$ is a path from x to y (that is, $\alpha(0) = x$ and $\alpha(1) = y$), then

$$\operatorname{length}\alpha \geqslant |x - y|.$$

Set

$$\|x - y\| = \inf\{\operatorname{length}\alpha\}$$

where the greatest lower bound is taken for all paths from x to y. It is straightforward to check that $(x, y) \mapsto \|x - y\|$ is an ∞-metric; moreover, $(\mathcal{X}, \|{*} - {*}\|)$ is a length space. The metric $\|{*} - {*}\|$ is called the induced length metric.

1.25. Exercise. Let $\mathcal{X}$ be a complete length space. Show that for any compact subset $K \subset \mathcal{X}$ There is a compact path-connected subset $K' \subset \mathcal{X}$ that contains K.

1.26. Exercise. Suppose $(\mathcal{X}, |{*} - {*}|)$ is a complete metric space. Show that $(\mathcal{X}, \|{*} - {*}\|)$ is complete.

Let A be a subset of a metric space $\mathcal{X}$. Given two points $x, y \in A$, consider the value

$$|x - y|_A = \inf_{\alpha}\{\operatorname{length}\alpha\},$$

where the greatest lower bound is taken for all paths α from x to y in A. In other words, $|{*} - {*}|_A$ denotes the induced length metric on the subspace A. (The notation

$|{*}-{*}|_A$ conflicts with the previously defined notation for distance $|x-y|_{\mathcal{X}}$ in a metric space $\mathcal{X}$. However, most of the time we will work with ambient length spaces where the meaning will be unambiguous.)

Let x and y be points in a metric space $\mathcal{X}$:

(i) A point $z \in \mathcal{X}$ is called a midpoint between x and y if

$$|x-z| = |y-z| = \tfrac{1}{2} \cdot |x-y|.$$

(ii) Assume $\varepsilon \geqslant 0$. A point $z \in \mathcal{X}$ is called an ε-midpoint between x and y if

$$|x-z| \leqslant \tfrac{1}{2} \cdot |x-y| + \varepsilon \quad \text{and} \quad |y-z| \leqslant \tfrac{1}{2} \cdot |x-y| + \varepsilon.$$

Note that a 0-midpoint is the same as a midpoint.

1.27. Menger's lemma. *Assume $\mathcal{X}$ is a complete metric space:*

(a) Suppose that for any two points in $\mathcal{X}$, and any positive ε, there is an ε-midpoint. Then $\mathcal{X}$ is a length space.
(b) Suppose that for any two points in $\mathcal{X}$, there is a midpoint. Then $\mathcal{X}$ is a geodesic space.

The second part of this lemma was proved by Karl Menger [43, Section 6]:

Proof.

(a) Choose $x, y \in \mathcal{X}$; set $\varepsilon_n = \frac{\varepsilon}{4^n}$, $\alpha(0) = x$, and $\alpha(1) = y$.

Let $\alpha(\frac{1}{2})$ be an ε_1-midpoint between $\alpha(0)$ and $\alpha(1)$. Further, let $\alpha(\frac{1}{4})$ and $\alpha(\frac{3}{4})$ be ε_2-midpoints between the pairs $(\alpha(0), \alpha(\frac{1}{2}))$ and $(\alpha(\frac{1}{2}), \alpha(1))$, respectively. Continue the above procedure; on the n-th step, we define $\alpha(\frac{k}{2^n})$, for every odd integer k such that $0 < \frac{k}{2^n} < 1$, as an ε_n-midpoint of the already defined $\alpha(\frac{k-1}{2^n})$ and $\alpha(\frac{k+1}{2^n})$.

This way we define $\alpha(t)$ for all dyadic rationals t in $[0,1]$. Moreover, α has Lipschitz constant $|x-y|+\varepsilon$. Since $\mathcal{X}$ is complete, the map α can be extended $(|x-y|+\varepsilon)$-Lipschitz map $\alpha\colon [0,1] \to \mathcal{X}$. In particular,

$$\text{③} \qquad \operatorname{length} \alpha \leqslant |x-y| + \varepsilon.$$

Since $\varepsilon > 0$ is arbitrary, we get *(a)*.

(b) Apply the same argument with midpoints instead of ε_n-midpoints. In this case, ③ holds for $\varepsilon_n = \varepsilon = 0$. □

In a compact space, a sequence of $\frac{1}{n}$-midpoints z_n contains a convergent subsequence. Therefore, Menger's Lemma (1.27) implies the following.

1.28. Proposition. *Any proper length space is geodesic.*

1.29. Hopf–Rinow theorem. *Any complete, locally compact length space is proper.*

Before reading the proof, it is instructive to solve 1.16. In the proof, we will use the following exercise.

1.30. Exercise. Let $\mathcal{X}$ be a length space. Show that $\mathrm{B}(x, R+\varepsilon)_{\mathcal{X}}$ is the ε-neighborhood of $\mathrm{B}(x, R)_{\mathcal{X}}$.

Proof. Choose a point x in a locally compact length space $\mathcal{X}$. Let

$$\rho(x) := \sup\left\{\, R \,:\, \overline{\mathrm{B}}[x, R] \text{ is compact} \,\right\}.$$

Since $\mathcal{X}$ is locally compact,

④ $$\rho(x) > 0 \quad \text{for any} \quad x \in \mathcal{X}.$$

It is sufficient to show that $\rho(x) = \infty$ for some (and therefore any) point $x \in \mathcal{X}$.

⑤ *If $\rho(x) < \infty$, then $B = \overline{\mathrm{B}}[x, \rho(x)]$ is compact.*

Suppose $\rho(x) > \varepsilon > 0$; by 1.30, the set $\overline{\mathrm{B}}[x, \rho(x) - \varepsilon]$ is a compact $2{\cdot}\varepsilon$-net in B. Since B is closed and hence complete, it must be compact; see 1.11(c) and 1.13. △

$|\rho(x) - \rho(y)| \leqslant |x - y|_{\mathcal{X}}$ for any $x, y \in \mathcal{X}$; in particular,

⑥ *$\rho\colon \mathcal{X} \to \mathbb{R}$ is a continuous function.*

Suppose $\rho(x) + |x - y| < \rho(y)$ for some $x, y \in \mathcal{X}$. Then $\overline{\mathrm{B}}[x, \rho(x) + \varepsilon]$ is a closed subset of $\overline{\mathrm{B}}[y, \rho(y)]$ for some $\varepsilon > 0$. Since $\overline{\mathrm{B}}[y, \rho(y)]$ is compact, so is $\overline{\mathrm{B}}[x, \rho(x) + \varepsilon]$—a contradiction. △

Let $\varepsilon = \min\{\, \rho(y) \,:\, y \in B \,\}$; the minimum is defined since B is compact and ρ is continuous. By ④, we have $\varepsilon > 0$.

Choose a finite $\frac{\varepsilon}{10}$-net $\{a_1, a_2, \dots, a_n\}$ in $B = \overline{\mathrm{B}}[x, \rho(x)]$. The union W of the closed balls $\overline{\mathrm{B}}[a_i, \varepsilon]$ is compact. By 1.30, $\overline{\mathrm{B}}[x, \rho(x) + \frac{\varepsilon}{10}] \subset W$. Therefore, $\overline{\mathrm{B}}[x, \rho(x) + \frac{\varepsilon}{10}]$ is compact, a contradiction. □

1.31. Exercise. Construct a geodesic space $\mathcal{X}$ that is locally compact, but whose completion $\bar{\mathcal{X}}$ is neither geodesic nor locally compact.

1.32. Advanced exercise. Show that for any compact connected space $\mathcal{X}$ there is a number ℓ such that for any finite collection of points there is a point z that lies on average distance ℓ from the collection; that is, for any $x_1, \dots, x_n \in \mathcal{X}$ there is $z \in \mathcal{X}$ such that

$$\tfrac{1}{n} \cdot \sum_i |x_i - z|_{\mathcal{X}} = \ell.$$

Lecture 2
Universal Spaces

The Urysohn space is the main hero of this lecture. It shares some fundamental properties with classical spaces (spheres, Euclidean, and Lobachevsky spaces) but also has many counterintuitive properties.

This space often serves as a counterexample to plausible conjectures; so it is worth knowing it. In addition, this space is beautiful.

A Embedding in a normed space

Recall that a function $v \mapsto |v|$ on a vector space $\mathcal{V}$ is called norm if it satisfies the following condition for any two vectors $v, w \in \mathcal{V}$ and a scalar α:

- $|v| \geqslant 0$.
- $|\alpha \cdot v| = |\alpha| \cdot |v|$.
- $|v| + |w| \geqslant |v + w|$.

As an example, consider ℓ^∞—the space of real sequences equipped with sup-norm; that is, the norm of $\boldsymbol{a} = (a_1, a_2, \dots)$ is defined by

$$|\boldsymbol{a}|_{\ell^\infty} := \sup_n \{\, |a_n| \,\}.$$

It is straightforward to check that for any normed space the function $(v, w) \mapsto |v - w|$ defines a metric on it. Therefore, any normed space is an example of metric space; moreover, it is a geodesic space. Often we do not distinguish normed space from the corresponding metric space. (By the Mazur–Ulam theorem, the metric remembers the affine structure of the space; so, to recover the original normed space, we only need to specify the origin. A slick proof of this theorem was given by Jussi Väisälä [65].)

A. Petrunin, *Pure Metric Geometry*, SpringerBriefs in Mathematics,
https://doi.org/10.1007/978-3-031-39162-0_2

Recall that diameter of a metric space $\mathcal{X}$ (briefly diam $\mathcal{X}$) is defined as the least upper bound on the distances between pairs of its points; that is,

$$\operatorname{diam} \mathcal{X} := \sup \left\{ |x - y|_{\mathcal{X}} : x, y \in \mathcal{X} \right\}.$$

If diam $\mathcal{X} < \infty$, then the space $\mathcal{X}$ is called bounded.

2.1. Lemma. *Suppose $\mathcal{X}$ is a bounded separable metric space; that is, $\mathcal{X}$ contains a countable, dense set, say $\{w_n\}$. Given $x \in \mathcal{X}$, set $a_n(x) = |w_n - x|_{\mathcal{X}}$. Then*

$$\iota \colon x \mapsto (a_1(x), a_2(x), \dots)$$

defines a distance-preserving embedding $\iota \colon \mathcal{X} \hookrightarrow \ell^\infty$.

Proof By the triangle inequality

$$① \qquad |a_n(x) - a_n(y)| \leqslant |x - y|_{\mathcal{X}}.$$

Therefore, ι is short (in other words, ι is distance-nonexpanding).

Again by the triangle inequality, we have

$$|a_n(x) - a_n(y)| \geqslant |x - y|_{\mathcal{X}} - 2 \cdot |w_n - x|_{\mathcal{X}}.$$

Since the set $\{w_n\}$ is dense, we can choose w_n arbitrarily close to x. Whence

$$② \qquad \sup_n \{ |a_n(x) - a_n(y)| \} \geqslant |x - y|_{\mathcal{X}};$$

that is, ι is distance-noncontracting.

Finally, observe that ① and ② imply the lemma. □

2.2. Exercise. Show that any compact metric space $\mathcal{K}$ is isometric to a subspace of a compact geodesic space.

The following exercise generalizes the lemma to arbitrary separable spaces.

2.3. Exercise. Suppose $\{w_n\}$ is a countable, dense set in a metric space $\mathcal{X}$. Choose $x_0 \in \mathcal{X}$; given $x \in \mathcal{X}$, set

$$a_n(x) = |w_n - x|_{\mathcal{X}} - |w_n - x_0|_{\mathcal{X}}.$$

Show that $\iota \colon x \mapsto (a_1(x), a_2(x), \dots)$ defines a distance-preserving embedding $\iota \colon \mathcal{X} \hookrightarrow \ell^\infty$.

Conclude that any separable metric space $\mathcal{X}$ admits a distance-preserving embedding $\iota \colon \mathcal{X} \hookrightarrow \ell^\infty$.

The following lemma implies that *any metric space is isometric to a subset of a normed vector space*; its proof is nearly identical to the proof of 2.3. Given a set $\mathcal{X}$, denote by $\ell^\infty(\mathcal{X})$ the space of all bounded functions on $\mathcal{X}$ equipped with sup-norm; that is,

$$|f - g|_{\ell^\infty} = \sup\{\,|f(x) - f(x)\; : \; x \in \mathcal{X}\,\}.$$

2.4. Lemma. *Let x_0 be a point in a metric space $\mathcal{X}$. Then the map $\iota\colon \mathcal{X} \to \ell^\infty(\mathcal{X})$ defined by*

$$\iota\colon x \mapsto (\mathrm{dist}_x - \mathrm{dist}_{x_0})$$

is distance-preserving.

In particular, any metric space $\mathcal{X}$ admits a distance-preserving into $\ell^\infty(\mathcal{X})$.

B Extension property

If a metric space $\mathcal{X}$ is a subspace of a semimetric space $\mathcal{X}'$, then we say that $\mathcal{X}'$ is an extension of $\mathcal{X}$. If in addition, $\operatorname{diam}\mathcal{X}' \leqslant d$, then we say that $\mathcal{X}'$ is a d-extension.

If the complement $\mathcal{X}' \setminus \mathcal{X}$ contains a single point, say p, then $\mathcal{X}'$ is called a one point extension of $\mathcal{X}$. In this case, to define a metric on $\mathcal{X}'$, it is sufficient to specify the distance function from p; that is, a function $f : \mathcal{X} \to \mathbb{R}$ defined by

$$f(x) := |p - x|_{\mathcal{X}'}.$$

Any function f of that type will be called an extension function or d-extension function, respectively.

The extension function f cannot be taken arbitrarily—the triangle inequality implies that

$$f(x) + f(y) \geqslant |x - y|_{\mathcal{X}} \geqslant |f(x) - f(y)|$$

for any $x, y \in \mathcal{X}$. In particular, f is a nonnegative 1-Lipschitz function on $\mathcal{X}$. For a d-extension, we need to assume in addition that $\operatorname{diam}\mathcal{X} \leqslant d$ and $f(x) \leqslant d$ for any $x \in \mathcal{X}$. A straightforward check shows that these conditions are necessary and sufficient.

2.5. Exercise. Let $\mathcal{X}$ be a subspace of metric space $\mathcal{Y}$. Assume f is an extension function on $\mathcal{X}$:

(a) Show that

$$\bar{f}(y) := \inf_{x \in \mathcal{X}}\{\, f(x) + |x - y|_{\mathcal{Y}}\,\}$$

defines an extension function on $\mathcal{Y}$.

(b) Assume that $\operatorname{diam}\mathcal{Y} \leqslant d$ and $f(x) \leqslant d$ for any $x \in \mathcal{X}$. Show that

$$\bar{f}_d := \min\{\bar{f}, d\}$$

is a d-extension function on $\mathcal{Y}$.

The functions $\bar{f}$ and $\bar{f}_d$ in the above exercise are called Katětov extensions of f, and the minimal possible $\mathcal{X}$ is called its support, briefly $\operatorname{supp}\bar{f} = \mathcal{X}$.

2.6. Definition. A metric space $\mathcal{U}$ meets the extension property if for any finite subspace $\mathcal{F} \subset \mathcal{U}$ and any extension function $f : \mathcal{F} \to \mathbb{R}$ there is a point $p \in \mathcal{U}$ such that $|p - x| = f(x)$ for any $x \in \mathcal{F}$.

If we assume in addition that $\operatorname{diam}\mathcal{U} \leqslant d$ and instead of extension functions we consider only d-extension functions, then it defines the d-extension property.

Further, if, in addition, $\mathcal{U}$ is separable and complete, then it is called Urysohn space or d-Urysohn space, respectively.

2.7. Proposition. *There is a separable metric space with the (d-) extension property (for any $d \geqslant 0$).*

Proof Choose $d \geqslant 0$. Let us construct a separable metric space with the d-extension property.

Let $\mathcal{X}$ be a metric space such that $\operatorname{diam}\mathcal{X} \leqslant d$. Denote by $\mathcal{X}^d$ the space of all d-extension functions on $\mathcal{X}$ equipped with the metric defined by the sup-norm. Note that the map $\mathcal{X} \to \mathcal{X}^d$ defined by $x \mapsto \operatorname{dist}_x$ is a distance-preserving embedding, so we can (and will) treat $\mathcal{X}$ as a subspace of $\mathcal{X}^d$; equivalently, $\mathcal{X}^d$ is an extension of $\mathcal{X}$.

Let us iterate this construction. Start with a one-point space $\mathcal{X}_0$ and consider a sequence of spaces $(\mathcal{X}_n)$ defined by $\mathcal{X}_{n+1}{:=}\mathcal{X}_n^d$. Note that the sequence is nested; that is, $\mathcal{X}_0 \subset \mathcal{X}_1 \subset \dots$ and the union

$$\mathcal{X}_\infty = \bigcup_n \mathcal{X}_n;$$

comes with metric such that $|x - y|_{\mathcal{X}_\infty} = |x - y|_{\mathcal{X}_n}$ if $x, y \in \mathcal{X}_n$.

Note that if $\mathcal{X}$ is compact, then so is $\mathcal{X}^d$. It follows that each space $\mathcal{X}_n$ is compact. In particular, $\mathcal{X}_\infty$ is a countable union of compact spaces; therefore, $\mathcal{X}_\infty$ is separable.

Any finite subspace $\mathcal{F}$ of $\mathcal{X}_\infty$ lies in some $\mathcal{X}_n$ for $n < \infty$. By construction, given an extension function $f : \mathcal{F} \to \mathbb{R}$, there is a point $p \in \mathcal{X}_{n+1}$ that meets the condition in 2.6. That is, $\mathcal{X}_\infty$ has the d-extension property.

The construction of a separable metric space with the extension property requires only two changes. First, the sequence should be defined by $\mathcal{X}_{n+1}{:=}\mathcal{X}_n^{d_n}$, where d_n is an increasing sequence such that $d_n \to \infty$. Second, the point p should be taken in $\mathcal{X}_{n+k}$ for sufficiently large k, so that $d_{n+k} > \max\{f(x)\}$ (here one has to apply 2.5(a)).

(Alternatively, one can start with any separable space $\mathcal{X}_0$ and consider a nested sequence $\mathcal{X}_0 \subset \mathcal{X}_1 \subset \dots$ where $\mathcal{X}_{n+1}$ is the space of all extension functions on $\mathcal{X}_n$ with at most $n+1$ points in its support. The last condition is needed to keep $\mathcal{X}_n$ separable.) □

Given a metric space $\mathcal{X}$, denote by $\mathcal{X}^\infty$ the space of all extension functions on $\mathcal{X}$ equipped with the metric defined by the sup-norm.

2.8. Exercise. Construct a proper length space $\mathcal{X}$ such that $\mathcal{X}^\infty$ is not separable.

2.9. Proposition. *If a metric space $\mathcal{V}$ meets the (d-) extension property, then so does its completion.*

Proof Let us assume $\mathcal{V}$ meets the extension property. We will show that its completion $\mathcal{U} = \bar{\mathcal{V}}$ meets the extension property as well. The d-extension case can be proved along the same lines.

Note that $\mathcal{V}$ is a dense subset in a complete space $\mathcal{U}$. Observe that $\mathcal{U}$ has the approximate extension property; that is, if $\mathcal{F} \subset \mathcal{U}$ is a finite set, $\varepsilon > 0$, and $f\colon \mathcal{F} \to \mathbb{R}$ is an extension function, then there exists $p \in \mathcal{U}$ such that

$$|p - x| \lesssim f(x) \pm \varepsilon \qquad ③$$

for any $x \in \mathcal{F}$. Indeed, consider the Katětov extension $\bar{f}\colon \mathcal{U} \to \mathbb{R}$ of f. Since $\mathcal{V}$ is dense in $\mathcal{U}$, we can choose a finite set $\mathcal{F}' \in \mathcal{V}$ such that for any $x \in \mathcal{F}$ there is $x' \in \mathcal{F}'$ with $|x - x'| < \frac{\varepsilon}{2}$. Let p be the point provided by the extension property for the restriction $\bar{f}|_{\mathcal{F}'}$. It remains to observe p meets ③.

It follows that there is a sequence of points $p_n \in \mathcal{U}$ such that for any $x \in \mathcal{F}$,

$$|p_n - x| \lesssim f(x) \pm \tfrac{1}{2^n}.$$

Moreover, we can assume that

$$|p_n - p_{n+1}| < \tfrac{1}{2^n} \qquad ④$$

for all large n. Indeed, consider the sets $\mathcal{F}_n = \mathcal{F} \cup \{p_n\}$ and the functions $f_n\colon \mathcal{F}_n \to \mathbb{R}$ defined by $f_n(x) := f(x)$ and

$$f_n(p_n) := \max\left\{\,\big||p_n - x| - f(x)\big| \;:\; x \in \mathcal{F}\right\}$$

if $x \neq p_n$. Observe that f_n is an extension function for large n and $f_n(p_n) < \frac{1}{2^n}$. Therefore, applying the approximate extension property recursively, we get ④.

Therefore, the sequence p_n is Cauchy. Note that its limit meets the condition in the definition of extension property (2.6). □

Note that 2.7 and 2.9 imply the following:

2.10. Theorem. *Urysohn space and d-Urysohn space exist for any $d > 0$.*

Here is a slightly stronger statement:

2.11. Theorem. *Any separable metric space $\mathcal{X}$ admits a distance-preserving embedding into an Urysohn space $\mathcal{U}$ such that any isometry of $\mathcal{X}$ can be extended to an isometry of $\mathcal{U}$.*

Sketch of proof Start with $\mathcal{X}_0 = \mathcal{X}$ and construct a nested sequence of spaces $\mathcal{X}_0 \subset \mathcal{X}_1 \subset \dots$ as at the alternative end of the proof of 2.7. Note that any isometry $\mathcal{X}_n \to \mathcal{X}_n$ can be extended to a unique isometry $\mathcal{X}_{n+1} \to \mathcal{X}_{n+1}$. It follows that any isometry of $\mathcal{X}$ can be extended to an isometry of $\mathcal{X}' = \bigcup_n \mathcal{X}_n$.

Now, consider a new nested sequence $\mathcal{X} \subset \mathcal{X}' \subset \mathcal{X}'' \subset \dots$; denote its union by $\mathcal{Y}$. Arguing as in 2.7 and 2.9, we get that the completion of $\mathcal{Y}$ is an Urysohn space, say $\mathcal{U}$, that comes with a distance-preserving inclusion $\mathcal{X} \hookrightarrow \mathcal{U}$.

From above, any isometry of $\mathcal{X}$ can be extended to isometries of $\mathcal{X}'$, $\mathcal{X}''$, and so on. They all define an isometry of $\mathcal{Y}$; passing to its continuous extension, we get an isometry of $\mathcal{U}$. □

C Universality

A metric space will be called universal if it has a subspace isometric to any given separable metric space. In 2.3, we proved that ℓ^∞ is a universal space. The following proposition shows that an Urysohn space is universal as well. Unlike ℓ^∞, Urysohn spaces are separable; so it might be considered as a *better* universal space. Theorem 2.20 will give another reason why Urysohn spaces are better.

2.12. Proposition. *An Urysohn space is universal. That is, if $\mathcal{U}$ is an Urysohn space, then any separable metric space $\mathcal{S}$ admits a distance-preserving embedding $\mathcal{S} \hookrightarrow \mathcal{U}$.*

Moreover, for any finite subspace $\mathcal{F} \subset \mathcal{S}$, any distance-preserving embedding $\mathcal{F} \hookrightarrow \mathcal{U}$ can be extended to a distance-preserving embedding $\mathcal{S} \hookrightarrow \mathcal{U}$.

A d-Urysohn space is d-universal; that is, the above statements hold provided that $\operatorname{diam} \mathcal{S} \leqslant d$.

Proof We will prove the second statement; the first statement is its partial case for $\mathcal{F} = \varnothing$.

The required isometry will be denoted by $x \mapsto x'$.

Choose a dense sequence of points $s_1, s_2, \dots \in \mathcal{S}$. We may assume that $\mathcal{F} = \{s_1, \dots, s_n\}$, so $s_i' \in \mathcal{U}$ are defined for $i \leqslant n$.

The sequence s_i' for $i > n$ can be defined recursively using the extension property in $\mathcal{U}$. Namely, suppose that $s_1', \dots, s_{i-1}'$ are already defined. Since $\mathcal{U}$ meets the extension property, there is a point $s_i' \in \mathcal{U}$ such that

$$|s_i' - s_j'|_{\mathcal{U}} = |s_i - s_j|_{\mathcal{S}}$$

for any $j < i$.

The constructed map $s_i \mapsto s_i'$ is distance-preserving. Therefore, it can be continuously extended to the whole $\mathcal{S}$. It remains to observe that the constructed map $\mathcal{S} \hookrightarrow \mathcal{U}$ is distance-preserving. □

2.13. Exercise. Show that any two distinct points in an Urysohn space can be joined by an infinite number of distinct geodesics.

2.14. Exercise. Modify the proofs of 2.9 and 2.12 to prove the following theorem.

2.15. Theorem. *Let K be a compact set in a separable space $\mathcal{S}$. Then any distance-preserving map from K to an Urysohn space can be extended to a distance-preserving map of the whole $\mathcal{S}$.*

2.16. Exercise. Show that (d-) Urysohn space is simply-connected.

D Uniqueness and homogeneity

2.17. Theorem. *Suppose $\mathcal{F} \subset \mathcal{U}$ and $\mathcal{F}' \subset \mathcal{U}'$ be finite isometric subspaces in a pair of (d-)Urysohn spaces $\mathcal{U}$ and $\mathcal{U}'$. Then any isometry $\iota\colon \mathcal{F} \leftrightarrow \mathcal{F}'$ can be extended to an isometry $\mathcal{U} \leftrightarrow \mathcal{U}'$.*

In particular, (d-)Urysohn space is unique up to isometry.

Note that 2.12 implies that there are distance-preserving maps $\mathcal{U} \to \mathcal{U}'$ and $\mathcal{U}' \to \mathcal{U}$. The next exercise shows that it does not solely imply the existence of an isometry $\mathcal{U} \leftrightarrow \mathcal{U}'$.

2.18. Exercise. Construct two metric spaces $\mathcal{X}$ and $\mathcal{Y}$ such that there are distance-preserving maps $\mathcal{X} \to \mathcal{Y}$ and $\mathcal{Y} \to \mathcal{X}$, but no isometry $\mathcal{X} \leftrightarrow \mathcal{Y}$.

The following construction uses the idea of 2.12, but it is applied back-and-forth to ensure that the obtained distance-preserving map is onto.

Proof Choose dense sequences $a_1, a_2, \dots \in \mathcal{U}$ and $b_1', b_2', \dots \in \mathcal{U}'$. We can assume that $\mathcal{F} = \{a_1, \dots, a_n\}$, $\mathcal{F}' = \{b_1', \dots, b_n'\}$, and $\iota(a_i) = b_i'$ for $i \leqslant n$.

The required isometry $\mathcal{U} \leftrightarrow \mathcal{U}'$ will be denoted by $u \leftrightarrow u'$. Set $a_i = b_i$ and $a_i' = b_i'$ if $i \leqslant n$.

Let us define recursively $a_{n+1}', b_{n+1}, a_{n+2}', b_{n+2}, \dots$—on the odd step, we define the images of $a_{n+1}, a_{n+2}, \dots$, and on the even steps, we define inverse images of $b_{n+1}', b_{n+2}', \dots$ The same argument as in the proof of 2.12 shows that we can construct two sequences $a_1', a_2', \dots \in \mathcal{U}'$ and $b_1, b_2, \dots \in \mathcal{U}$ such that

$$|a_i - a_j|_{\mathcal{U}} = |a_i' - a_j'|_{\mathcal{U}'},$$

$$|a_i - b_j|_{\mathcal{U}} = |a_i' - b_j'|_{\mathcal{U}'},$$

$$|b_i - b_j|_{\mathcal{U}} = |b_i' - b_j'|_{\mathcal{U}'}$$

for all i and j.

It remains to observe that the constructed distance-preserving bijection defined by $a_i \leftrightarrow a_i'$ and $b_i \leftrightarrow b_i'$ extends continuously to an isometry $\mathcal{U} \leftrightarrow \mathcal{U}'$. □

Observe that 2.17 implies that the Urysohn space (as well as the d-Urysohn space) is finite-set-homogeneous; that is:

- Any distance-preserving map from a finite subset to the whole space can be extended to an isometry.

Recall that $S(p,r)_{\mathcal{X}}$ denotes the sphere of radius r centered at p in a metric space $\mathcal{X}$; that is,

$$S(p,r)_{\mathcal{X}} = \left\{ x \in \mathcal{X} \,:\, |p - x|_{\mathcal{X}} = r \right\}.$$

2.19. Exercise. Choose $d \in [0, \infty]$. Denote by $\mathcal{U}_d$ the d-Urysohn space, so $\mathcal{U}_\infty$ is the Urysohn space:

(a) Assume that $L = S(p,r)_{\mathcal{U}_d} \neq \varnothing$. Show that L is isometric to $\mathcal{U}_\ell$; find ℓ in terms of r and d.
(b) Let $\ell = |p - q|_{\mathcal{U}_d}$. Show that the subset $M \subset \mathcal{U}_d$ of midpoints between p and q is isometric to $\mathcal{U}_\ell$.
(c) Show that $\mathcal{U}_d$ is not countable-set-homogeneous; that is, there is a distance-preserving map from a countable subset of $\mathcal{U}_d$ to $\mathcal{U}_d$ that cannot be extended to an isometry of $\mathcal{U}_d$.

In fact, the Urysohn space is compact-set-homogeneous; more precisely, the following theorem holds.

2.20. Theorem. *Let K be a compact set in a (d-)Urysohn space $\mathcal{U}$. Then any distance-preserving map $K \to \mathcal{U}$ can be extended to an isometry of $\mathcal{U}$.*

A proof can be obtained by modifying the proofs of 2.9 and 2.17 the same way as it is done in 2.14.

2.21. Exercise. Let S be a unit sphere in the Urysohn space $\mathcal{U}$. Show that for any two distinct points $x, y \in \mathcal{U}$ there is a point $z \in S$ such that $|x - z| \neq |y - z|$.

Conclude that two isometries of $\mathcal{U}$ coincide if they coincide on S.

2.22. Exercise. Let B be an open unit ball in the Urysohn space $\mathcal{U}$. Show that $\mathcal{U} \setminus B$ is isometric to $\mathcal{U}$.

Use it to construct an isometry of a unit sphere S in $\mathcal{U}$ that cannot be extended to an isometry of $\mathcal{U}$.

2.23. Exercise.

(a) Show that there is a distance-preserving inclusion of the Urysohn space $\iota\colon \mathcal{U} \hookrightarrow \mathcal{U}$ such that $\mathcal{U}' = \iota(\mathcal{U})$ is nowhere dense in $\mathcal{U}$ and any isometry of $\mathcal{U}'$ can be extended to an isometry of the whole $\mathcal{U}$.
(b) Consider a nested sequence $\mathcal{U}_0 \subset \mathcal{U}_1 \subset \dots$ of Urysohn spaces with each inclusion $\mathcal{U}_n \hookrightarrow \mathcal{U}_{n+1}$ as in (a). Show that the union $\bigcup_n \mathcal{U}_n$ is a noncomplete finite-set-homogeneous metric space that meets the extension property.

2.24. Exercise. Which of the following metric spaces are one-point-homogeneous, finite-set-homogeneous, compact-set-homogeneous, countable-set-homogeneous?

(a) Euclidean plane
(b) Hilbert space ℓ^2
(c) ℓ^∞
(d) ℓ^1—the space of all real absolutely converging series $\boldsymbol{a} = (a_1, a_2, \dots)$ with the norm $|\boldsymbol{a}|_{\ell^1} = \sum_i |a_i|$

2.25. Exercise. Show that any separable one-point-homogeneous metric tree is isometric to the real line $\mathbb{R}$ or the one-point space.

E Remarks

The statement in 2.3 was proved by Maurice René Fréchet in the paper where he first defined metric spaces [19]; its extension 2.4 was given by Kazimierz Kuratowski [37]. Both maps $x \mapsto (\mathrm{dist}_x - \mathrm{dist}_{x_0})$ and $x \mapsto \mathrm{dist}_x$ can be called Kuratowski embedding.

Let us describe a closely related construction introduced by Mikhael Gromov [21]. Suppose $\mathcal{X}$ be a proper metric space. Denote by $C(\mathcal{X}, \mathbb{R})$ the space of continuous functions $\mathcal{X} \to \mathbb{R}$ equipped with the compact-open topology; that is, for any compact set $K \subset \mathcal{X}$ and any open set $U \subset \mathbb{R}$ the set of all continuous functions $f : \mathcal{X} \to \mathbb{R}$ such that $f(K) \subset U$ is declared to be open.

Choose a point $x_0 \in \mathcal{X}$. Consider the map $F_\mathcal{X} : \mathcal{X} \to C(\mathcal{X}, \mathbb{R})$ defined by

$$x \mapsto f_x := -|x - x_0| + \mathrm{dist}_x.$$

2.26. Exercise. Show that if $\mathcal{X}$ is a proper length space, then $F_\mathcal{X}$ is an embedding. Construct a proper metric space $\mathcal{Y}$ such that $F_\mathcal{Y}$ is not an embedding.

Denote by $\hat{\mathcal{X}}$ the closure of $F_\mathcal{X}(\mathcal{X})$ in $C(\mathcal{X}, \mathbb{R})$; observe that $\hat{\mathcal{X}}$ is compact. If $F_\mathcal{X}$ is an embedding, then $\hat{\mathcal{X}}$ is a compactification of $\mathcal{X}$, and it is called the horo-compactification. In this case, the complement

$$\partial_\infty \mathcal{X} = \hat{\mathcal{X}} \setminus F_\mathcal{X}(\mathcal{X})$$

is called the horo-absolute of $\mathcal{X}$. A variation of this construction for nonproper spaces was considered by Anders Karlsson [33].

The following two exercises show that in this respect ℓ^∞ is very different from ℓ^1. For more on the subject, see [16].

Let S be a subset of X. We say that S separates x and y if $x \in S$ and $y \notin S$ or $x \notin S$ and $y \in S$. The cut metric δ_S on X is a semimetric such that $\delta_S(x, y) = 1$ if x and y are separated by S and otherwise $\delta_S(x, y) = 0$.

2.27. Exercise. Show that a finite metric space $\mathcal{F}$ admits a distance-preserving embedding into ℓ^1 if and only if the metric of $\mathcal{F}$ can be written as a nonnegative linear combination [1] of cut metrics on $\mathcal{F}$.

Recall that the vertex set of any graph comes with the path metric—the distance between two vertices is the minimal number of edges in a path connecting them.

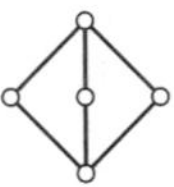

2.28. Exercise. Use 2.27 to show that the metric for complete bipartite graph $K_{2,3}$ (see the diagram) does not admit a distance-preserving embedding into ℓ^1.

The question about the existence of a separable universal space was posed by Maurice René Fréchet and answered by Pavel Urysohn [63]. Exercise 2.23 answers a question posed by Pavel Urysohn [63, §2(6)]. It was solved by Miroslav Katětov [34], but long after that, it was again mentioned as an open problem [22, p. 83].

The idea of Urysohn's construction was reused in graph theory; it produces the so-called Rado graph, also known as *Erdős–Rényi graph* or *random graph*; see [15]. In fact, the Urysohn space is the random metric space in *certain sense* [66].

The (d-)Urysohn space is homeomorphic to the Hilbert space; the latter was proved by Vladimir Uspenskij [64] using the so-called Toruńczyk criterion.

The finite-set-homogeneous spaces include Euclidean spaces, hyperbolic spaces, and spheres all with standard length metrics and arbitrary finite dimensions. In fact, these are the only examples of locally compact three-point-homogeneous length spaces. The latter was proved by Herbert Busemann [14]; it also follows from the more general result of Jacques Tits about two-point-homogeneous spaces [62]. The same conclusion holds for complete all-set-homogeneous geodesic spaces with local uniqueness of geodesics; it was proved by Garrett Birkhoff [8]. The answer might be the same for complete separable all-set-homogeneous length spaces. Without the separability condition, we also get the so-called universal metric trees with finite valence [17]; no other examples seem to be known [40].

2.29. Exercise. Show that the real projective plane $\mathbb{R}\mathrm{P}^2$ with the standard metric is two-point-homogeneous, but not three-point-homogeneous.

2.30. Exercise. Let Q be the set of vertices on the n-dimensional cube; assume n is large. Show that Q is three-point-homogeneous, but not four-point-homogeneous.

2.31. Question. Are there examples of metric spaces that are n-point-homogeneous, but not $(n+1)$-point-homogeneous for large n? See [51].

[1] That is, linear combination with nonnegative coefficients.

Lecture 3
Injective Spaces

Injective hull is a useful construction that provides a canonical choice of a specially nice (injective) space that includes a given metric space. This construction is similar to the convex hull in Euclidean space. The following exercise gives a bridge from the latter to the former.

3.1. Advanced exercise. Show that $A \subset \mathbb{R}^n$ is a closed convex set if and only if for any $B \subset \mathbb{R}^n$ any short map $B \to A$ can be extended to a short map $\mathbb{R}^n \to A$.

A Definition

3.2. Definition. A metric space $\mathcal{Y}$ is called injective if, for any metric space $\mathcal{X}$ and any of its subspace $\mathcal{A}$, any short map $f: \mathcal{A} \to \mathcal{Y}$ can be extended to a short map $F: \mathcal{X} \to \mathcal{Y}$; that is, $f = F|_{\mathcal{A}}$.

3.3. Exercise. Show that any injective space is:

(a) Complete (b) Geodesic (c) Contractible

3.4. Exercise. Show that for any injective space $\mathcal{Y}$ there is a map $m: \mathcal{Y} \times \mathcal{Y} \to \mathcal{Y}$ (the midpoint map) such that the inequality

$$2 \cdot |p - m(x, y)|_{\mathcal{Y}} \leqslant |p - x|_{\mathcal{Y}} + |p - y|_{\mathcal{Y}}$$

holds for any $p, x, y \in \mathcal{Y}$.

3.5. Exercise. Show that the following spaces are injective:

(a) The real line.
(b) Complete metric tree.

A. Petrunin, *Pure Metric Geometry*, SpringerBriefs in Mathematics,
https://doi.org/10.1007/978-3-031-39162-0_3

(c) The space $\ell^\infty(\mathcal{S})$ for any set $\mathcal{S}$ (defined in 2.4). In particular, the coordinate plane with the metric induced by the ℓ^∞-norm.

3.6. Exercise. Let $\mathcal{Y}$ be an injective space:

(a) Show that any closed ball in $\mathcal{Y}$ is injective.
(b) Show that the intersection of an arbitrary collection of closed balls in $\mathcal{Y}$ is injective.

3.7. Advanced exercise. Let $\mathcal{Y}$ be a bounded injective space. Show that any short map $s\colon \mathcal{Y} \to \mathcal{Y}$ has a fixed point.

B Admissible and extremal functions

Let $\mathcal{X}$ be a metric space. A function $r\colon \mathcal{X} \to (-\infty, \infty]$ is called admissible if the following inequality

$$r(x) + r(y) \geqslant |x - y|_{\mathcal{X}} \qquad ①$$

holds for any $x, y \in \mathcal{X}$.

3.8. Observation.

(a) Any admissible function is nonnegative.
(b) If $\mathcal{X}$ is a geodesic space, then a function $r\colon \mathcal{X} \to \mathbb{R}$ is admissible if and only if

$$\overline{\mathrm{B}}[x, r(x)] \cap \overline{\mathrm{B}}[y, r(y)] \neq \varnothing$$

for any $x, y \in \mathcal{X}$.

Proof

(a) Apply ① for $x = y$.
(b) Apply the triangle inequality and the existence of a geodesic $[xy]$. □

A minimal admissible function will be called extremal. More precisely, an admissible function $r\colon \mathcal{X} \to \mathbb{R}$ is extremal if for any admissible function $s\colon \mathcal{X} \to \mathbb{R}$ we have

$$s \leqslant r \quad \Longrightarrow \quad s = r.$$

Applying Zorn's lemma, we get the following.

3.9. Observation. *For any admissible function s, there is an extremal function r such that $r \leqslant s$.*

3.10. Lemma. *Let r be an extremal function and s an admissible function on a metric space $\mathcal{X}$. Suppose that $r \geqslant s - c$ for some constant c. Then $r \leqslant s + c$; in particular, $c \geqslant 0$.*

Proof Note that if $c < 0$, then $r > s$. The latter is impossible since r is extremal and s is admissible.

Observe that the function $\bar{r} = \min\{r, s + c\}$ is admissible. Indeed, choose $x, y \in \mathcal{X}$. If $\bar{r}(x) = r(x)$ and $\bar{r}(y) = r(y)$, then

$$\bar{r}(x) + \bar{r}(y) = r(x) + r(y) \geqslant |x - y|.$$

Further, if $\bar{r}(x) = s(x) + c$, then

$$\begin{aligned}\bar{r}(x) + \bar{r}(y) &\geqslant [s(x) + c] + [s(y) - c] = \\ &= s(x) + s(y) \geqslant \\ &\geqslant |x - y|.\end{aligned}$$

Since r is extremal, we have $r = \bar{r}$; that is, $r \leqslant s + c$. □

3.11. Observations. *Let $\mathcal{X}$ be a metric space:*

(a) For any point $p \in \mathcal{X}$, the distance function $r = \operatorname{dist}_p$ is extremal.
(b) Any extremal function r on $\mathcal{X}$ is 1-Lipschitz; that is,

$$|r(p) - r(q)| \leqslant |p - q|$$

for any $p, q \in \mathcal{X}$. In other words, any extremal function is an extension function [see Sect. B].
(c) An admissible function r on $\mathcal{X}$ is extremal if and only if for any point $p \in \mathcal{X}$ and any $\delta > 0$, there is a point $q \in \mathcal{X}$ such that

$$r(p) + r(q) < |p - q|_{\mathcal{X}} + \delta.$$

(d) Suppose $\mathcal{X}$ is compact. Then an admissible function r on $\mathcal{X}$ is extremal if and only if for any point $p \in \mathcal{X}$ there is a point $q \in \mathcal{X}$ such that

$$r(p) + r(q) = |p - q|_{\mathcal{X}}.$$

Proof

(a) By the triangle inequality, ① holds; that is, $r = \operatorname{dist}_p$ is an admissible function.
Further, if $s \leqslant r$ is another admissible function, then $s(p) = 0$ and ① implies that $s(x) \geqslant |p - x|$. Whence $s = r$.

(b) By *(a)*, dist_p is admissible. Since r is admissible, we have that

$$r \geqslant \operatorname{dist}_p - r(p).$$

Since r is extremal, 3.10 implies that

$$r \leqslant \operatorname{dist}_p + r(p),$$

or, equivalently,

$$r(q) - r(p) \leqslant |p - q|$$

for any $p, q \in \mathcal{X}$. Whence the statement follows.

(c) Assume r is extremal. Arguing by contradiction, assume there is $\delta > 0$ such that

$$r(q) \geqslant \operatorname{dist}_p(q) - r(p) + \delta$$

for any q. By *(a)*, dist_p is extremal; in particular, admissible. Therefore, 3.10 implies that

$$r(q) \leqslant \operatorname{dist}_p(q) + r(p) - \delta$$

for any q. Taking $q = p$, we get $r(p) \leqslant r(p) - \delta$, a contradiction.

Now suppose r is not extremal; that is, there is an admissible function $s \leqslant r$ such that $r(p) - s(p) = \delta > 0$ for some p. Then, for any q, we have

$$r(p) + r(q) \geqslant s(p) + s(q) + \delta \geqslant |p - q|_{\mathcal{X}} + \delta$$

—a contradiction.

(d) The if part follows from *(c)*.

Denote by q_n the point provided by *(c)* for $\delta = \frac{1}{n}$. Let q be a partial limit of q_n. Then

$$r(p) + r(q) \leqslant |p - q|_{\mathcal{X}}.$$

Since r is admissible, the opposite inequality holds; whence the only-if part follows.

□

3.12. Exercise. Consider the unit circle

$$\mathbb{S}^1 = \left\{ (x, y) : x^2 + y^2 = 1 \right\}$$

in the plane with induced length metric. Show that $r\colon \mathbb{S}^1 \to \mathbb{R}$ is extremal if and only if it is 1-Lipschitz and

$$r(p) + r(-p) = \pi$$

for any $p \in \mathbb{S}^1$.

3.13. Exercise. Given a real-valued function s on a metric space $\mathcal{X}$, consider the function

$$s^*(x) = \sup\left\{\, |x-y|_{\mathcal{X}} - s(y) \;:\; y \in \mathcal{X} \,\right\}.$$

Show that the function $\tfrac{1}{2}\cdot(s+s^*)$ is admissible for any s.

C Equivalent conditions

3.14. Theorem. *For any metric space $\mathcal{Y}$, the following conditions are equivalent:*

(a) $\mathcal{Y}$ is injective.
(b) If $r\colon \mathcal{Y}\to\mathbb{R}$ is an extremal function, then there is a point $p\in\mathcal{Y}$ such that

$$|p-x| = r(x)$$

for any $x\in\mathcal{Y}$.
(c) $\mathcal{Y}$ is hyperconvex; that is, if $\left\{\overline{\mathrm{B}}[x_\alpha, r_\alpha] \;:\; \alpha\in\mathcal{A}\right\}$ is a family of closed balls in $\mathcal{Y}$ such that

$$r_\alpha + r_\beta \geqslant |x_\alpha - x_\beta|$$

for any $\alpha,\beta\in\mathcal{A}$, then all the balls in the family $\{\overline{\mathrm{B}}[x_\alpha, r_\alpha]\}_{\alpha\in\mathcal{A}}$ have a common point.

Proof We will prove implications *(a)⇒(b)⇒(c)⇒(a)*.
(a)⇒(b). By 3.11(b), r is an extension function. Applying the definition of injective space to a one-point extension of $\mathcal{Y}$, we get a point $p\in\mathcal{Y}$ such that

$$|p-x| = \mathrm{dist}_p(x) \leqslant r(x)$$

for any $x\in\mathcal{Y}$. By 3.11(a), the distance function dist_p is extremal. Since r is extremal, we get $\mathrm{dist}_p = r$.
(b)⇒(c). By 3.8(b), part *(c)* is equivalent to the following statement:

- If $r: \mathcal{Y} \to \mathbb{R}$ is an admissible function, then there is a point $p \in \mathcal{Y}$ such that

②
$$|p - x| \leqslant r(x)$$

for any $x \in \mathcal{Y}$.

Indeed, set $r(x) := \inf\{\, r_\alpha : x_\alpha = x \,\}$. (If $x_\alpha \neq x$ for any α, then $r(x) = \infty$.) The condition in *(c)* implies that r is admissible. It remains to observe that $p \in \overline{\mathrm{B}}[x_\alpha, r_\alpha]$ for every α if and only if ② holds.

By 3.9, for any admissible function r, there is an extremal function $\bar{r} \leqslant r$; hence, *(b)*⇒*(c)*.

(c)⇒*(a)*. Arguing by contradiction, suppose $\mathcal{Y}$ is not injective; that is, there is a metric space $\mathcal{X}$ with a subset $\mathcal{A}$ such that a short map $f: \mathcal{A} \to \mathcal{Y}$ cannot be extended to a short map $F: \mathcal{X} \to \mathcal{Y}$. By Zorn's lemma, we may assume that $\mathcal{A}$ is a maximal subset; that is, the domain of f cannot be enlarged by a single point.[1]

Fix a point p in the complement $\mathcal{X} \setminus \mathcal{A}$. To extend f to p, we need to choose $f(p)$ in the intersection of the balls $\overline{\mathrm{B}}[f(x), r(x)]$, where $r(x) = |p - x|$. Therefore, this intersection for all $x \in \mathcal{A}$ has to be empty.

Since f is short, we have that

$$\begin{aligned} r(x) + r(y) &\geqslant |x - y|_{\mathcal{X}} \geqslant \\ &\geqslant |f(x) - f(y)|_{\mathcal{Y}}. \end{aligned}$$

By *(c)*, the balls $\overline{\mathrm{B}}[f(x), r(x)]$ have a common point—a contradiction. □

3.15. Exercise. Suppose a length space $\mathcal{W}$ has two subspaces $\mathcal{X}$ and $\mathcal{Y}$ such that $\mathcal{X} \cup \mathcal{Y} = \mathcal{W}$ and $\mathcal{X} \cap \mathcal{Y}$ is a one-point set. Assume $\mathcal{X}$ and $\mathcal{Y}$ are injective. Show that $\mathcal{W}$ is injective.

3.16. Exercise. Show that an m-dimensional normed space is injective if and only if it is isometric to $\mathbb{R}^m$ with ℓ^∞-norm; that is,

$$|(x_1, \ldots, x_m)| = \max_i \{\, |x_i| \,\}.$$

A metric space $\mathcal{Y}$ is called finitely hyperconvex or countably hyperconvex if the condition in 3.14(*c*) holds only for any finite or respectively countable family of balls.

3.17. Exercise. Show that any proper finitely hyperconvex metric space is hyperconvex.

3.18. Exercise. Show that the d-Urysohn space is finitely hyperconvex, but not countably hyperconvex. Conclude that the d-Urysohn space is not injective.

[1] In this case, $\mathcal{A}$ must be closed, but we will not use it.

Try to do the same for the Urysohn space.

3.19. Exercise. Let $\mathcal{Y}$ be a complete metric space. Suppose $\mathcal{Y}$ is almost hyperconvex, that is, for any $\varepsilon > 0$ any family of closed balls $\{\overline{B}[x_\alpha, r_\alpha + \varepsilon] : \alpha \in \mathcal{A}\}$ has a common point if $r_\alpha + r_\beta \geqslant |x_\alpha - x_\beta|$ for all $\alpha, \beta \in \mathcal{A}$. Show that $\mathcal{Y}$ is hyperconvex.

D Space of extremal functions

Let $\mathcal{X}$ be a metric space. Consider the space $\operatorname{Ext}\mathcal{X}$ of extremal functions on $\mathcal{X}$ equipped with sup-norm; that is,

$$|f - g|_{\operatorname{Ext}\mathcal{X}} := \sup\{\,|f(x) - g(x)| : x \in \mathcal{X}\,\}.$$

Recall that by 3.11(*a*), any distance function is extremal. It follows that the map $x \mapsto \operatorname{dist}_x$ produces a distance-preserving embedding $\mathcal{X} \hookrightarrow \operatorname{Ext}\mathcal{X}$. So we can (and will) treat $\mathcal{X}$ as a subspace of $\operatorname{Ext}\mathcal{X}$, or, equivalently, $\operatorname{Ext}\mathcal{X}$ as an extension of $\mathcal{X}$. In particular, from now on, a point $x \in \mathcal{X}$ can refer to the function $\operatorname{dist}_x : \mathcal{X} \to \mathbb{R}$ and the other way around.

Since any extremal function is 1-Lipschitz, for any $f \in \operatorname{Ext}\mathcal{X}$ and $p \in \mathcal{X}$, we have that $f(x) \leqslant f(p) + \operatorname{dist}_p(x)$. By 3.10, we also get $f(x) \geqslant -f(p) + \operatorname{dist}_p(x)$. Therefore,

$$\text{③}\qquad \begin{aligned}|f - p|_{\operatorname{Ext}\mathcal{X}} &= \sup\{\,|f(x) - \operatorname{dist}_p(x)| : x \in \mathcal{X}\,\} = \\ &= f(p).\end{aligned}$$

In particular, the statement in 3.11(*c*) can be written as

$$|f - p|_{\operatorname{Ext}\mathcal{X}} + |f - q|_{\operatorname{Ext}\mathcal{X}} < |p - q|_{\operatorname{Ext}\mathcal{X}} + \delta.$$

3.20. Exercise. Show that $\operatorname{Ext}\mathcal{X}$ is compact if and only if so is $\mathcal{X}$.

3.21. Exercise. Describe the set of all extremal functions on a metric space $\mathcal{X}$ and the metric space $\operatorname{Ext}\mathcal{X}$ in each of the following cases:

(a) $\mathcal{X}$ is a metric space with exactly two points v, w on distance 1 from each other.
(b) $\mathcal{X}$ is a metric space with exactly three points a, b, c such that

$$|a - b|_{\mathcal{X}} = |b - c|_{\mathcal{X}} = |c - a|_{\mathcal{X}} = 1.$$

(c) $\mathcal{X}$ is a metric space with exactly four points p, q, x, y such that

$$|p - x|_{\mathcal{X}} = |p - y|_{\mathcal{X}} = |q - x|_{\mathcal{X}} = |q - y|_{\mathcal{X}} = 1$$

and

$$|p - q|_{\mathcal{X}} = |x - y|_{\mathcal{X}} = 2.$$

3.22. Exercise. Assume $\mathcal{X}$ is a compact metric space. Recall that the map $x \mapsto \mathrm{dist}_x$ gives an isometric embedding $\mathcal{X} \hookrightarrow \ell^\infty(\mathcal{X})$; so we can think that $\mathcal{X}$ is a subset of $\ell^\infty(\mathcal{X})$.

Given two points $x, y \in \mathcal{X}$, denote by $G_{x,y}$ the union of all geodesics from x to y in $\ell^\infty(\mathcal{X})$. Show that $\mathrm{Ext}\,\mathcal{X}$ is isometric to

$$G = \bigcap_{x\in\mathcal{X}} \left(\bigcup_{y\in\mathcal{X}} G_{x,y} \right).$$

3.23. Proposition. $\mathrm{Ext}\,\mathcal{X}$ *is injective for any metric space* $\mathcal{X}$.

3.24. Lemma. *Let* $\mathcal{X}$ *be a metric space. Then*

$$\sigma \in \mathrm{Ext}\,(\mathrm{Ext}\,\mathcal{X}) \quad \Longrightarrow \quad \sigma|_{\mathcal{X}} \in \mathrm{Ext}\,\mathcal{X}.$$

In other words, if σ is an extremal function on $\mathrm{Ext}\,\mathcal{X}$, then the restriction of σ to $\mathcal{X}$ is an extremal function on $\mathcal{X}$.

Proof Arguing by contradiction, suppose that there is an admissible function $s\colon \mathcal{X} \to \mathbb{R}$ such that $s(x) \leqslant \sigma(x)$ for any $x \in \mathcal{X}$ and $s(p) < \sigma(p)$ for some point $p \in \mathcal{X}$. Consider another function $\bar\sigma : \mathrm{Ext}\,\mathcal{X} \to \mathbb{R}$ such that $\bar\sigma(f) := \sigma(f)$ if $f \neq p$ and $\bar\sigma(p) := s(p)$.

Let us show that $\bar\sigma$ is admissible; that is,

④ $$|f - g|_{\mathrm{Ext}\,\mathcal{X}} \leqslant \bar\sigma(f) + \bar\sigma(g)$$

for any $f, g \in \mathrm{Ext}\,\mathcal{X}$.

Since σ is admissible and $\bar\sigma = \sigma$ on $(\mathrm{Ext}\,\mathcal{X}) \setminus \{p\}$, it is sufficient to prove 4 assuming $f \neq g = p$. By ③, we have $|f - p|_{\mathrm{Ext}\,\mathcal{X}} = f(p)$. Therefore, 4 boils down to the following inequality

⑤ $$\sigma(f) + s(p) \geqslant f(p),$$

for any $f \in \mathrm{Ext}\,\mathcal{X}$.

Fix small $\delta > 0$. Let $q \in \mathcal{X}$ be the point provided by 3.11*(c)*. Then

$$\sigma(f) + s(p) \geqslant [\sigma(f) - \sigma(q)] + [\sigma(q) + s(p)] \geqslant$$

since σ is 1-Lipschitz, and $\sigma(q) \geqslant s(q)$, we can continue

$$\geqslant -|q - f|_{\mathrm{Ext}\,\mathcal{X}} + [s(q) + s(p)] \geqslant$$

by ③ and since s is admissible

$$\geqslant -f(q) + |p - q| >$$

and by 3.11*(c)*

$$> f(p) - \delta.$$

Since $\delta > 0$ is arbitrary, ⑤ and ④ follow.

Summarizing: the function $\bar{\sigma}$ is admissible, $\bar{\sigma} \leqslant \sigma$ and $\bar{\sigma}(p) < \sigma(p)$; that is, σ is not extremal—a contradiction. □

Proof of 3.23 Choose a function $\sigma \in \operatorname{Ext}(\operatorname{Ext}\mathcal{X})$. By 3.24, $s{:=}\sigma|_{\mathcal{X}} \in \operatorname{Ext}\mathcal{X}$; that is, s is extremal. By 3.14(*b*), it is sufficient to show that

⑥ $$\sigma(f) \geqslant |s - f|_{\operatorname{Ext}\mathcal{X}}$$

for any $f \in \operatorname{Ext}\mathcal{X}$.

Since σ is 1-Lipschitz (3.11(*b*)), we have that

$$s(x) - f(x) = \sigma(x) - |f - x|_{\operatorname{Ext}\mathcal{X}} \leqslant \sigma(f),$$

for any $x \in \mathcal{X}$. By 3.10, $s(x) - f(x) \geqslant -\sigma(f)$ for any $x \in \mathcal{X}$. Whence ⑥ follows. □

3.25. Exercise. Let $\mathcal{X}$ be a compact metric space. Show that for any two points $f, g \in \operatorname{Ext}\mathcal{X}$ lie on a geodesic $[pq]$ with $p, q \in \mathcal{X}$.

A metric space $\mathcal{X}$ is called δ-hyperbolic if

$$|p - q| + |x - y| \leqslant \max\{\,|p - x| + |q - y|,\ |p - y| + |q - x|\,\} + 2\cdot\delta$$

for any $p, q, x, y \in \mathcal{X}$.

3.26. Advanced exercise. Show that $\operatorname{Ext}\mathcal{X}$ is δ-hyperbolic if and only if so is $\mathcal{X}$.

E Injective envelope

An extension $\mathcal{E}$ of a metric space $\mathcal{X}$ will be called its injective envelope if $\mathcal{E}$ is an injective space, and there is no proper injective subspace of $\mathcal{E}$ that contains $\mathcal{X}$.

Two injective envelopes $e\colon \mathcal{X} \hookrightarrow \mathcal{E}$ and $f\colon \mathcal{X} \hookrightarrow \mathcal{F}$ are called equivalent if there is an isometry $\iota\colon \mathcal{E} \to \mathcal{F}$ such that $f = \iota \circ e$.

3.27. Theorem. *For any metric space $\mathcal{X}$, its extension* $\operatorname{Ext}\mathcal{X}$ *is an injective envelope.*

Moreover, any other injective envelope of $\mathcal{X}$ is equivalent to $\operatorname{Ext}\mathcal{X}$.

Proof Suppose $S \subset \operatorname{Ext}\mathcal{X}$ is an injective subspace containing $\mathcal{X}$. Since S is injective, there is a short map $w\colon \operatorname{Ext}\mathcal{X} \to S$ that fixes all points in $\mathcal{X}$.

Suppose that $w\colon f \mapsto f'$; observe that $f(x) \geqslant f'(x)$ for any $x \in \mathcal{X}$. Since f is extremal, $f = f'$; that is, w is the identity map, and therefore, $S = \operatorname{Ext}\mathcal{X}$.

Assume we have another injective envelope $e\colon \mathcal{X} \hookrightarrow \mathcal{E}$. Then there are short maps $v\colon \mathcal{E} \to \operatorname{Ext}\mathcal{X}$ and $w\colon \operatorname{Ext}\mathcal{X} \to \mathcal{E}$ such that $x = v \circ e(x)$ and $e(x) = w(x)$ for any $x \in \mathcal{X}$. From above, the composition $v \circ w$ is the identity on $\operatorname{Ext}\mathcal{X}$. In particular, w is distance-preserving.

The composition $w \circ v\colon \mathcal{E} \to \mathcal{E}$ is a short map that fixes points in $e(\mathcal{X})$. Since $e\colon \mathcal{X} \hookrightarrow \mathcal{E}$ is an injective envelope, the composition $w \circ v$ and therefore w are onto. Whence w is an isometry. □

3.28. Exercise. Suppose $e\colon \mathcal{X} \hookrightarrow \mathcal{E}$ and $f\colon \mathcal{X} \hookrightarrow \mathcal{F}$ are two injective envelopes of $\mathcal{X}$. Show that there is a unique isometry $\iota\colon \mathcal{E} \to \mathcal{F}$ such that $\iota \circ e = f$.

3.29. Exercise. Suppose $\mathcal{X}$ is a subspace of a metric space $\mathcal{U}$. Show that the inclusion $\mathcal{X} \hookrightarrow \mathcal{U}$ can be extended to a distance-preserving inclusion $\operatorname{Ext}\mathcal{X} \hookrightarrow \operatorname{Ext}\mathcal{U}$.

3.30. Exercise. Consider the hemisphere

$$\mathbb{S}^2_+ = \left\{ (x, y, z) \in \mathbb{R}^3 \ : \ x^2 + y^2 + z^2 = 1, \quad z \geqslant 0 \right\}$$

and its boundary

$$\mathbb{S}^1 = \left\{ (x, y, z) \in \mathbb{R}^3 \ : \ x^2 + y^2 + z^2 = 1, \quad z = 0 \right\},$$

both with induced length metrics.

Show that there is unique isometric embedding $\iota\colon \mathbb{S}^2_+ \hookrightarrow \operatorname{Ext}\mathbb{S}^1$ such that $\iota(u) = u$ for any $u \in \mathbb{S}^1$.

F Remarks

Injective spaces were introduced by Nachman Aronszajn and Prom Panitchpakdi [3]. The injective envelope was introduced by John Isbell [28]; it is also known as tight span and hyperconvex hull.

It was observed by John Isbell [29] that *if $\mathcal{X}$ is a Banach space, then its injective hull* $\operatorname{Ext}\mathcal{X}$ *has a natural structure of Banach space* (which is unique by the Mazur–Ulam theorem). Moreover, $\mathcal{X}$ is a linear subspace of $\operatorname{Ext}\mathcal{X}$.

Let us mention that a metric space $\mathcal{X}$ is called convex if for any pair of points $x_1, x_2 \in \mathcal{X}$ and any $r_1, r_2 \geqslant 0$, we have

$$r_1 + r_2 \geqslant |x_1 - x_2|_{\mathcal{X}} \qquad \Longrightarrow \qquad \overline{\mathrm{B}}[x, r_1]_{\mathcal{X}} \cap \overline{\mathrm{B}}[y, r_2]_{\mathcal{X}} \neq \varnothing;$$

in other words, a pair of balls intersect if the triangle inequality does not forbid it. Clearly, hyperconvexity (3.14(*c*)) is stronger than convexity. Note that *any geodesic space is convex*. The converse does not hold in general, but by Menger's lemma (1.27(*b*)) *any complete convex space is geodesic*.

More generally, a metric space $\mathcal{X}$ is called n-hyperconvex if the condition in 3.14(*c*) holds only for families with at most n balls; so *convex means* 2-*hyperconvex*.

The following striking result was proved by Benjamin Miesch and Maël Pavón [46].

3.31. Theorem. *Any complete* 4-*hyperconvex space is finitely hyperconvex.*

So, by 3.17, it follows that *any proper* 4-*hyperconvex space is hyperconvex.*

3.32. Exercise. Show that ℓ^1 is 3-but not 4-hyperconvex.

Recall that if the following inequality:

$$|x - z|_{\mathcal{X}} \leqslant \max\{\,|x - y|_{\mathcal{X}}, |y - z|_{\mathcal{X}}\,\}$$

holds for any three points x, y, z in a metric space $\mathcal{X}$, then $\mathcal{X}$ is called an ultrametric space. In some sense, ultrametric spaces are dual to injective spaces.

3.33. Exercise. Suppose that a metric space $\mathcal{X}$ satisfies the following property: For any subspace $\mathcal{A}$ in $\mathcal{X}$ and any other metric space $\mathcal{Y}$, any short map $f: \mathcal{A} \to \mathcal{Y}$ can be extended to a short map $F: \mathcal{X} \to \mathcal{Y}$.

Show that $\mathcal{X}$ is an ultrametric space.

A subspace $\mathcal{S}$ of a metric space $\mathcal{X}$ is called its short retract if there is a short map $\mathcal{X} \to \mathcal{S}$ that is the identity on $\mathcal{S}$.

3.34. Exercise. Show that any compact subspace $\mathcal{K}$ of an ultrametric space $\mathcal{X}$ is its short retract.

Construct an example of a complete ultrametric space $\mathcal{X}$ with a closed subspace $\mathcal{Q}$ that is not its short retract.

The following exercise gives a sufficient condition for the existence of a short extension.

3.35. Exercise. Let $f: A \to \mathcal{K}$ be a short map from a subset A in a metric space $\mathcal{X}$ to compact metric space $\mathcal{K}$. Assume that for any finite set $F \subset \mathcal{X}$ there is a short map $F \to \mathcal{K}$ that agrees with f on $F \cap A$. Show that there is a short map $\mathcal{X} \to \mathcal{K}$ that agrees with f on A.

Lecture 4
Space of Subsets

In this lecture, we define and study Hausdorff metric on subsets of a given metric space.

A Hausdorff distance

Let $\mathcal{X}$ be a metric space. Given a subset $A \subset \mathcal{X}$, consider the distance function to A

$$\operatorname{dist}_A : \mathcal{X} \to [0, \infty)$$

defined as

$$\operatorname{dist}_A(x) := \inf\{\, |a - x|_{\mathcal{X}} \ :\ a \in A \,\}.$$

Further, we define the so-called Hausdorff metric on all nonempty compact subsets of a given metric space $\mathcal{X}$. The obtained metric space will be denoted as $\operatorname{Haus}\mathcal{X}$.

4.1. Definition. Let A and B be two nonempty compact subsets of a metric space $\mathcal{X}$. Then the Hausdorff distance between A and B is defined as

$$|A - B|_{\operatorname{Haus}\mathcal{X}} := \sup_{x \in \mathcal{X}}\{\, |\operatorname{dist}_A(x) - \operatorname{dist}_B(x)| \,\}.$$

The following observation gives a useful reformulation of the definition:

4.2. Observation. *Suppose A and B be two compact subsets of a metric space $\mathcal{X}$. Then $|A - B|_{\operatorname{Haus}\mathcal{X}} < R$ if and only if B lies in an R-neighborhood of A, and A lies in an R-neighborhood of B.*

A. Petrunin, *Pure Metric Geometry*, SpringerBriefs in Mathematics,
https://doi.org/10.1007/978-3-031-39162-0_4

4.3. Exercise. Let $\mathcal{X}$ be a metric space. Given a subset $A \subset \mathcal{X}$, define its diameter as

$$\operatorname{diam} A := \sup_{a,b \in A} |a - b|.$$

Show that

$$\operatorname{diam} : \operatorname{Haus} \mathcal{X} \to \mathbb{R}$$

is a 2-Lipschitz function; that is,

$$|\operatorname{diam} A - \operatorname{diam} B| \leqslant 2 \cdot |A - B|_{\operatorname{Haus} \mathcal{X}}$$

for any two compact nonempty sets $A, B \subset \mathcal{X}$.

4.4. Exercise. Let A and B be two compact subsets in the Euclidean plane $\mathbb{R}^2$. Assume $|A - B|_{\operatorname{Haus} \mathbb{R}^2} < \varepsilon$:

(a) Show that $|\operatorname{Conv} A - \operatorname{Conv} B|_{\operatorname{Haus} \mathbb{R}^2} < \varepsilon$, where $\operatorname{Conv} A$ denoted the convex hull of A.
(b) Is it true that $|\partial A - \partial B|_{\operatorname{Haus} \mathbb{R}^2} < \varepsilon$, where ∂A denotes the boundary of A. Does the converse hold? That is, assume A and B be two compact subsets in $\mathbb{R}^2$ and $|\partial A - \partial B|_{\operatorname{Haus} \mathbb{R}^2} < \varepsilon$; is it true that $|A - B|_{\operatorname{Haus} \mathbb{R}^2} < \varepsilon$?

Note that part (a) implies that $A \mapsto \operatorname{Conv} A$ defines a short map $\operatorname{Haus} \mathbb{R}^2 \to \operatorname{Haus} \mathbb{R}^2$.

4.5. Exercise. Let A and B be compact subsets in metric space $\mathcal{X}$. Show that

$$|A - B|_{\operatorname{Haus} \mathcal{X}} = \sup_f \{ \max_{a \in A}\{f(a)\} - \max_{b \in B}\{f(b)\} \},$$

where the least upper bound is taken for all 1-Lipschitz functions f.

Given a subset $A \subset \mathbb{R}^n$, the support function $h_A \colon \mathbb{R}^n \to \mathbb{R}$ of a nonempty closed set $A \subset \mathbb{R}^n$ is defined as

$$h_A(x) := \sup\{\, \langle x, a \rangle \;:\; a \in A \,\}.$$

4.6. Exercise. Show that

$$|A - B|_{\operatorname{Haus} \mathbb{R}^n} \geqslant \sup_{|u|=1} \{ |h_A(u) - h_B(u)| \}$$

for any nonempty compact subsets $A, B \subset \mathbb{R}^n$.

Moreover, equality holds if both A and B are convex.

4.7. Advanced exercise. Suppose $C_t \subset \mathcal{X}$, $t \in [0,1]$ is a family of subsets. A path $c\colon [0,1] \to \mathcal{X}$ such that $c(t) \in C_t$ for all t will be called a section of C_t:

(a) Construct a family of nonempty compact sets $C_t \subset \mathbb{S}^1$, $t \in [0,1]$ that is continuous in the Hausdorff topology but does not admit a section.
(b) Show that any family of nonempty compact sets $C_t \subset \mathbb{R}$, $t \in [0,1]$ that is continuous in the Hausdorff topology admits a section.

B Hausdorff convergence

4.8. Blaschke selection theorem. *A metric space $\mathcal{X}$ is compact if and only if so is* Haus $\mathcal{X}$.

The Hausdorff metric can be used to define convergence. Namely, suppose $K_1, K_2, \dots$ and K_∞ are compact sets in a metric space $\mathcal{X}$. If $|K_\infty - K_n|_{\mathrm{Haus}\,\mathcal{X}} \to 0$ as $n \to \infty$, then we say that the sequence K_n converges to K_∞ in the sense of Hausdorff; equivalently, K_∞ is the Hausdorff limit of the sequence K_n.

Note that the theorem implies that from any sequence of nonempty compact sets in $\mathcal{X}$ one can select a convergent subsequence; for that reason, it is called a *selection* theorem.

Proof; if part Consider the map ι that sends each point $x \in \mathcal{X}$ to the one point subset $\{x\}$ of $\mathcal{X}$. Note that $\iota\colon \mathcal{X} \to \mathrm{Haus}\,\mathcal{X}$ is distance-preserving.

Suppose that $A \subset \mathcal{X}$. Note that $\operatorname{diam} A = 0$ if and only if A is a one-point set. By 4.3, $\iota(\mathcal{X})$ is a closed subset of the compact space Haus $\mathcal{X}$. It follows that $\iota(\mathcal{X})$ and, therefore, $\mathcal{X}$ are compact. □

Since the map ι above is distance-preserving, we can and will consider $\mathcal{X}$ as a subspace of Haus $\mathcal{X}$.

4.9. Exercise. Let $\mathcal{X}$ be a compact length space. Suppose that there is a short retraction Haus $\mathcal{X} \to \mathcal{X}$. Show that $\mathcal{X}$ is contractible.

To prove the only-if part, we will need the following two lemmas.

4.10. Monotone convergence. *Let $K_1 \supset K_2 \supset \dots$ be a nested sequence of nonempty compact sets in a metric space $\mathcal{X}$. Then $K_\infty = \bigcap_n K_n$ is the Hausdorff limit of K_n; that is, $|K_\infty - K_n|_{\mathrm{Haus}\,\mathcal{X}} \to 0$ as $n \to \infty$.*

Proof By finite intersection property, K_∞ is a nonempty compact set.

Arguing by contradiction, assume that there is $\varepsilon > 0$ such that for each n one can choose $x_n \in K_n$ such that $\operatorname{dist}_{K_\infty}(x_n) \geqslant \varepsilon$. Note that $x_n \in K_1$ for each n. Since K_1 is compact, there is a partial limit x_∞ of x_n; that is, a limit of a subsequence. Clearly, $\operatorname{dist}_{K_\infty}(x_\infty) \geqslant \varepsilon$.

On the other hand, since K_n is closed and $x_m \in K_n$ for $m \geqslant n$, we get $x_\infty \in K_n$ for each n. It follows that $x_\infty \in K_\infty$ and therefore $\operatorname{dist}_{K_\infty}(x_\infty) = 0$— a contradiction. □

4.11. Lemma. *If $\mathcal{X}$ is a compact metric space, then* Haus $\mathcal{X}$ *is complete.*

Proof Let $Q_1, Q_2, \dots$ be a Cauchy sequence in Haus $\mathcal{X}$. Passing to a subsequence, we may assume that

① $$|Q_n - Q_{n+1}|_{\operatorname{Haus}\mathcal{X}} \leqslant \tfrac{1}{10^n}$$

for each n.

Denote by K_n the closed $\frac{2}{10^n}$-neighborhood of Q_n; that is,

$$K_n = \left\{ x \in \mathcal{X} \ : \ \operatorname{dist}_{Q_n}(x) \leqslant \tfrac{2}{10^n} \right\}.$$

Since $\mathcal{X}$ is compact so is each K_n.

From ①, we get $K_n \supset K_{n+1}$ for each n. Set

$$K_\infty = \bigcap_{n=1}^{\infty} K_n.$$

By the monotone convergence (4.10), $|K_n - K_\infty|_{\operatorname{Haus}\mathcal{X}} \to 0$ as $n \to \infty$.

By 4.2, $|Q_n - K_n|_{\operatorname{Haus}\mathcal{X}} \leqslant \frac{2}{10^n}$. Therefore, $|Q_n - K_\infty|_{\operatorname{Haus}\mathcal{X}} \to 0$ as $n \to \infty$—hence the lemma. □

4.12. Exercise. Let $\mathcal{X}$ be a complete metric space and $K_1, K_2, \dots$ be a sequence of compact sets that converges in the sense of Hausdorff. Show that the union $K_1 \cup K_2 \cup \dots$ has compact closure.

Use this statement to show that in Lemma 4.11 compactness of $\mathcal{X}$ can be exchanged to completeness.

Proof of only-if part in 4.8 According to Lemma 4.11, Haus $\mathcal{X}$ is complete. It remains to show that Haus $\mathcal{X}$ is totally bounded (1.11(c)); that is, given $\varepsilon > 0$, there is a finite ε-net in Haus $\mathcal{X}$.

Choose a finite ε-net A in $\mathcal{X}$. Denote by B the set of all nonempty subsets of A. Note that B is a finite set in Haus $\mathcal{X}$. For each compact set $K \subset \mathcal{X}$, consider the subset K' of all points $a \in A$ such that $\operatorname{dist}_K(a) \leqslant \varepsilon$. Observe that $K' \in B$ and $|K - K'|_{\operatorname{Haus}\mathcal{X}} \leqslant \varepsilon$. In other words, B is a finite ε-net in Haus $\mathcal{X}$. □

4.13. Exercise. Let $\mathcal{X}$ be a complete metric space. Show that $\mathcal{X}$ is a length space if and only if so is Haus $\mathcal{X}$:

4.14. Exercise.

(a) Show that the set of all connected compact subsets of $\mathbb{R}^2$ is closed in Haus $\mathbb{R}^2$.
(b) Show that any connected compact subset of $\mathbb{R}^2$ is a Hausdorff limit of a sequence of closed simple curves.

C An application

In this section, we will sketch a proof of the isoperimetric inequality in the plane that uses the Hausdorff convergence.

It is based on the following exercise.

4.15. Exercise. Let $\mathcal{C}$ be the set of all nonempty compact convex subsets in $\mathbb{R}^2$. Show that $\mathcal{C}$ is a closed subset of $\operatorname{Haus}\mathbb{R}^2$ and perimeter and area are continuous on $\mathcal{C}$. (If the set degenerates to a line segment of length ℓ, then its perimeter is defined as $2 \cdot \ell$.)

More precisely, if a sequence of convex compact plane sets X_n converges to X_∞ in the sense of Hausdorff, then X_∞ is convex,

$$\operatorname{perim} X_n \to \operatorname{perim} X_\infty, \quad \text{and} \quad \operatorname{area} X_n \to \operatorname{area} X_\infty$$

as $n \to \infty$.

4.16. Isoperimetric inequality. *Among the plane figures bounded by closed curves of length at most ℓ, the round disk has the maximal area.*

Sketch. It is sufficient to consider only convex figures of the given perimeter; if a figure is not convex, pass to its convex hull and observe that it has a larger area and smaller perimeter.

Note that the selection theorem (4.8) together with the exercise implies the existence of figure D with perimeter ℓ and maximal area.

It remains to show that D is a round disk; it will be done by means of elementary geometry.

Let us cut D along a chord $[ab]$ into two lenses, L_1 and L_2. Denote by L_1' the reflection of L_1 across the perpendicular bisector of $[ab]$. Note that D and $D' = L_1' \cup L_2$ have the same perimeter and area. That is, D' has perimeter ℓ and maximal possible area; in particular, D' is convex.

The following exercise will finish the proof. □

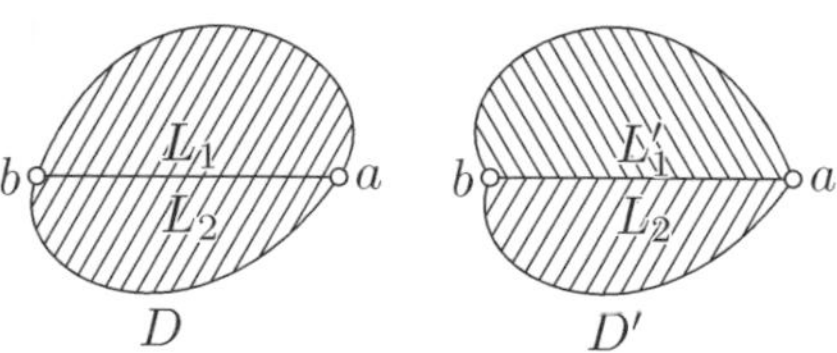

4.17. Exercise. Suppose D is a convex figure such that for any chord $[ab]$ of D the above construction produces a convex figure D'. Show that D is a round disk.

Another popular way to prove that D is a round disk is given by the so-called *Steiner's 4-joint method* [9].

D Remarks

It seems that Hausdorff convergence was first introduced by Felix Hausdorff [24]. A couple of years later an equivalent definition was given by Wilhelm Blaschke [9].

The following refinement was introduced by Zdeněk Frolík [20] and rediscovered by Robert Wijsman [67]. This refinement is also called Hausdorff convergence; in fact, it takes an intermediate place between the original Hausdorff convergence and the so-called *closed convergence*, also introduced by Hausdorff in [24].

4.18. Definition. Let $A_1, A_2, \dots$ be a sequence of closed sets in a metric space $\mathcal{X}$. We say that the sequence A_n converges to a closed set A_∞ in the sense of Hausdorff if, for any $x \in \mathcal{X}$, we have $\mathrm{dist}_{A_n}(x) \to \mathrm{dist}_{A_\infty}(x)$ as $n \to \infty$.

For example, suppose $\mathcal{X}$ is the Euclidean plane and A_n is the circle with radius n and center at the point $(0, n)$. If we use the standard definition (4.1), then the sequence $A_1, A_2, \dots$ diverges, but it converges to the x-axis in the sense of Definition 4.18.

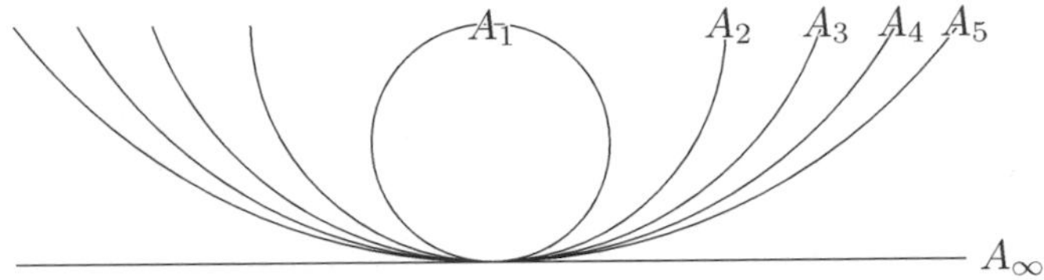

Further, consider the sequence of one-point sets $B_n = \{(n, 0)\}$ in the Euclidean plane. It diverges in the sense of the standard definition, but, in the sense of 4.18, it converges to the empty set; indeed, for any point x, we have $\mathrm{dist}_{B_n}(x) \to \infty$ as $n \to \infty$ and $\mathrm{dist}_n(x) = \infty$ for any x.

The following exercise is analogous to the Blaschke selection theorem (4.8) for the modified Hausdorff convergence.

4.19. Exercise. Let $\mathcal{X}$ be a proper metric space and $A_1, A_2, \dots$ be a sequence of closed sets in $\mathcal{X}$. Show that the sequence $A_1, A_2, \dots$ has a convergent subsequence in the sense of Definition 4.18.

Lecture 5
Space of Spaces

In this lecture, we define and study the so-called Gromov–Hausdorff metric on the isometry classes of compact metric spaces.

A Gromov–Hausdorff metric

In this section, we cook up a metric space out of all compact metric spaces. More precisely, we want to define the so-called Gromov–Hausdorff metric on the set of *isometry classes* of compact metric spaces. (Being isometric is an equivalence relation, and an isometry class is an equivalence class with respect to this relation.)

The obtained metric space will be denoted by GH. Given two metric spaces $\mathcal{X}$ and $\mathcal{Y}$, denote by $[\mathcal{X}]$ and $[\mathcal{Y}]$ their isometry classes; that is, $\mathcal{X}' \in [\mathcal{X}]$ if and only if $\mathcal{X}' \overset{iso}{=\!=} \mathcal{X}$. Pedantically, the Gromov–Hausdorff distance from $[\mathcal{X}]$ to $[\mathcal{Y}]$ should be denoted as $|[\mathcal{X}] - [\mathcal{Y}]|_{\mathrm{GH}}$; but we will write it as $|\mathcal{X} - \mathcal{Y}|_{\mathrm{GH}}$ and say (not quite correctly) that $|\mathcal{X} - \mathcal{Y}|_{\mathrm{GH}}$ *is the Gromov–Hausdorff distance from* $\mathcal{X}$ to $\mathcal{Y}$. In other words, from now on, the term *metric space* might also stand for its *isometry class*.

The metric on GH is defined as the maximal metric such that *the distance between subspaces in a metric space is not greater than the Hausdorff distance between them*. Here is a formal definition:

5.1. Definition. The Gromov–Hausdorff distance $|\mathcal{X}-\mathcal{Y}|_{\mathrm{GH}}$ between compact metric spaces $\mathcal{X}$ and $\mathcal{Y}$ is defined by the following relation.

Given $r > 0$, we have that $|\mathcal{X}-\mathcal{Y}|_{\mathrm{GH}} < r$ if and only if there exists a metric space $\mathcal{W}$ and subspaces $\mathcal{X}'$ and $\mathcal{Y}'$ in $\mathcal{W}$ that are isometric to $\mathcal{X}$ and $\mathcal{Y}$, respectively, such that $|\mathcal{X}' - \mathcal{Y}'|_{\mathrm{Haus}\,\mathcal{W}} < r$. (Here $|\mathcal{X}' - \mathcal{Y}'|_{\mathrm{Haus}\,\mathcal{W}}$ denotes the Hausdorff distance between sets $\mathcal{X}'$ and $\mathcal{Y}'$ in $\mathcal{W}$.)

A. Petrunin, *Pure Metric Geometry*, SpringerBriefs in Mathematics,
https://doi.org/10.1007/978-3-031-39162-0_5

5.2. Theorem. *The set of isometry classes of compact metric spaces equipped with Gromov–Hausdorff metric forms a metric space (which is denoted by* GH*).*

In other words, for arbitrary compact metric spaces $\mathcal{X}$, $\mathcal{Y}$, and $\mathcal{Z}$, the following conditions hold:

(a) $|\mathcal{X} - \mathcal{Y}|_{\mathrm{GH}} \geqslant 0$.
(b) $|\mathcal{X} - \mathcal{Y}|_{\mathrm{GH}} = 0$ *if and only if $\mathcal{X}$ is isometric to $\mathcal{Y}$.*
(c) $|\mathcal{X} - \mathcal{Y}|_{\mathrm{GH}} = |\mathcal{Y} - \mathcal{X}|_{\mathrm{GH}}$.
(d) $|\mathcal{X} - \mathcal{Y}|_{\mathrm{GH}} + |\mathcal{Y} - \mathcal{Z}|_{\mathrm{GH}} \geqslant |\mathcal{X} - \mathcal{Z}|_{\mathrm{GH}}$.

Note that *(a)*, *(c)*, and the if part of *(b)* follow directly from 5.1. Part *(d)* will be proved in Sect. B. The only-if part of *(b)* will be proved in Sect. C.

Recall that $a \cdot \mathcal{X}$ denotes $\mathcal{X}$ rescaled by a factor $a > 0$; that is, $a \cdot \mathcal{X}$ is a metric space with the underlying set of $\mathcal{X}$ and the metric defined by

$$|x - y|_{a\cdot\mathcal{X}} := a \cdot |x - y|_{\mathcal{X}}.$$

5.3. Exercise. Let $\mathcal{X}$ be a compact metric space, and $\mathcal{O}$ be the one-point metric space. Prove the following:

(a) $|\mathcal{X} - \mathcal{O}|_{\mathrm{GH}} = \frac{1}{2} \cdot \operatorname{diam} \mathcal{X}$.
(b) $|a \cdot \mathcal{X} - b \cdot \mathcal{X}|_{\mathrm{GH}} = \frac{1}{2} \cdot |a - b| \cdot \operatorname{diam} \mathcal{X}$.
(c) $\iota[\mathcal{O}] = [\mathcal{O}]$ for any isometry $\iota\colon \mathrm{GH} \to \mathrm{GH}$.

5.4. Exercise. Find two subsets $A, B \subset \mathbb{R}^2$ such that

$$|A - B|_{\mathrm{GH}} > |A - \iota(B)|_{\mathrm{Haus}\,\mathbb{R}^2}$$

for any isometry ι of $\mathbb{R}^2$.

5.5. Exercise. Let $\mathcal{A}_r$ be a rectangle 1 by r in the Euclidean plane and $\mathcal{B}_r$ be a closed line interval of length r. Show that

$$|\mathcal{A}_r - \mathcal{B}_r|_{\mathrm{GH}} > \tfrac{1}{10}$$

for all large r.

5.6. Advanced exercise. Let $\mathcal{X}$ and $\mathcal{Y}$ be compact metric spaces; denote by $\hat{\mathcal{X}}$ and $\hat{\mathcal{Y}}$ their injective envelopes (see Sect. D). Show that

$$|\hat{\mathcal{X}} - \hat{\mathcal{Y}}|_{\mathrm{GH}} \leqslant 2 \cdot |\mathcal{X} - \mathcal{Y}|_{\mathrm{GH}}.$$

In other words, $\mathcal{X} \mapsto \hat{\mathcal{X}}$ defines a 2-Lipschitz map $\mathrm{GH} \to \mathrm{GH}$.

B Approximations and almost isometries

5.7. Definition. Let $\mathcal{X}$ and $\mathcal{Y}$ be two metric spaces. A relation $\approx$ between points in $\mathcal{X}$ and $\mathcal{Y}$ is called ε-approximation if the following conditions hold:

- For any $x \in \mathcal{X}$, there is $y \in \mathcal{Y}$ such that $x \approx y$.
- For any $y \in \mathcal{Y}$, there is $x \in \mathcal{X}$ such that $x \approx y$.
- If $x \approx y$ and $x' \approx y'$ for some $x, x' \in \mathcal{X}$ and $y, y' \in \mathcal{Y}$, then

$$|x - x'|_{\mathcal{X}} \leq |y - y'|_{\mathcal{Y}}| \pm 2 \cdot \varepsilon.$$

5.8. Exercise. Let $\mathcal{X}$ and $\mathcal{Y}$ be two compact metric spaces. Show that

$$|\mathcal{X} - \mathcal{Y}|_{\mathrm{GH}} < \varepsilon$$

if and only if there is an ε-approximation between $\mathcal{X}$ and $\mathcal{Y}$.

In other words, $|\mathcal{X} - \mathcal{Y}|_{\mathrm{GH}}$ is the greatest lower bound of values $\varepsilon > 0$ such that there is an ε-approximation between $\mathcal{X}$ and $\mathcal{Y}$.

Proof of 5.2d Suppose that:

- $\approx_1$ is a relation between points in $\mathcal{X}$ and $\mathcal{Y}$
- $\approx_2$ is a relation between points in $\mathcal{Y}$ and $\mathcal{Z}$.

Consider the relation $\approx_3$ between points in $\mathcal{X}$ and $\mathcal{Z}$ such that $x \approx_3 z$ if and only if there is $y \in \mathcal{Y}$ such that $x \approx_1 y$ and $y \approx_2 z$.

It is straightforward to check that if $\approx_1$ is an ε_1-approximation and $\approx_2$ is an ε_2-approximation, then $\approx_3$ is an $(\varepsilon_1 + \varepsilon_2)$-approximation.

Applying 5.8, we get that if

$$|\mathcal{X} - \mathcal{Y}|_{\mathrm{GH}} < \varepsilon_1 \quad \text{and} \quad |\mathcal{Y} - \mathcal{Z}|_{\mathrm{GH}} < \varepsilon_2,$$

then

$$|\mathcal{X} - \mathcal{Z}|_{\mathrm{GH}} < \varepsilon_1 + \varepsilon_2.$$

Hence, 5.2(*d*) follows. □

The following weakened version of isometry is closely related to ε-approximations.

5.9. Definition. Let $\mathcal{X}$ and $\mathcal{Y}$ be metric spaces and $\varepsilon > 0$. A map[1] $f\colon \mathcal{X} \to \mathcal{Y}$ is called an ε-isometry if $f(\mathcal{X})$ is an ε-net in $\mathcal{Y}$ and

$$|x - x'|_{\mathcal{X}} \lesssim |f(x) - f(x')|_{\mathcal{Y}}| \pm \varepsilon$$

for any $x, x' \in \mathcal{X}$.

5.10. Exercise. Let $\mathcal{X}$ and $\mathcal{Y}$ be compact metric spaces:

(a) If $|\mathcal{X} - \mathcal{Y}|_{\mathrm{GH}} < \varepsilon$, then there is a $2 \cdot \varepsilon$-isometry $f\colon \mathcal{X} \to \mathcal{Y}$.
(b) If there is an ε-isometry $f\colon \mathcal{X} \to \mathcal{Y}$, then $|\mathcal{X} - \mathcal{Y}|_{\mathrm{GH}} < \varepsilon$.

C Optimal realization

Note that

$$|\mathcal{X}' - \mathcal{Y}'|_{\mathrm{Haus}\ \mathcal{W}} \geqslant |\mathcal{X} - \mathcal{Y}|_{\mathrm{GH}},$$

where $\mathcal{X}$, $\mathcal{Y}$, $\mathcal{X}'$, $\mathcal{Y}'$, and $\mathcal{W}$ are as in 5.1. The following proposition states that equality holds for some choice of $\mathcal{X}'$, $\mathcal{Y}'$, and $\mathcal{W}$.

5.11. Proposition. *For any two compact metric spaces $\mathcal{X}$ and $\mathcal{Y}$, there is a metric space $\mathcal{W}$ with subsets $\mathcal{X}'$ and $\mathcal{Y}'$ such that $\mathcal{X}' \overset{iso}{=\!=} \mathcal{X}$, $\mathcal{Y}' \overset{iso}{=\!=} \mathcal{Y}$, and*

$$|\mathcal{X}' - \mathcal{Y}'|_{\mathrm{Haus}\ \mathcal{W}} = |\mathcal{X} - \mathcal{Y}|_{\mathrm{GH}}.$$

Let us introduce the so-called *appropriate functions* and use them in a reinterpretation of the Gromov–Hausdorff distance.

Suppose $\mathcal{X}$, $\mathcal{Y}$, $\mathcal{X}'$, $\mathcal{Y}'$, and $\mathcal{W}$ are as in 5.1. By passing to the subspace $\mathcal{X}' \cup \mathcal{Y}'$ in $\mathcal{W}$, we can assume that $\mathcal{W} = \mathcal{X}' \cup \mathcal{Y}'$. Note that in this case the metric on $\mathcal{W}$ is completely determined by the function $f\colon \mathcal{X} \times \mathcal{Y} \to \mathbb{R}$ defined by

$$f(x, y) := |x - y|_{\mathcal{W}};$$

a function f that can appear this way will be called appropriate.

Note that a function $f\colon \mathcal{X} \times \mathcal{Y} \to \mathbb{R}$ is appropriate if and only if $x \mapsto f(x, y)$ and $y \mapsto f(x, y)$ are extension functions (see Sect. B); that is, if

① $$\begin{aligned} &f(x, y) + f(x, y') \geqslant |y - y'|_{\mathcal{Y}} \geqslant |f(x, y) - f(x, y')|, \quad \text{and} \\ &f(x, y) + f(x', y) \geqslant |x - x'|_{\mathcal{X}} \geqslant |f(x, y) - f(x', y)| \end{aligned}$$

[1] Possibly noncontinuous.

for any $x, x', \in \mathcal{X}$ and $y, y' \in \mathcal{X}$. In other words, the following defines a semimetric on $\mathcal{X} \sqcup \mathcal{Y}$

$$|x-y|_{\mathcal{X}\sqcup\mathcal{Y}} := \begin{cases} |x-y|_{\mathcal{X}} & \text{if } x, y \in \mathcal{X}, \\ |x-y|_{\mathcal{Y}} & \text{if } x, y \in \mathcal{Y}, \\ f(x,y) & \text{if } x \in \mathcal{X} \text{ and } y \in \mathcal{Y}, \end{cases}$$

and the corresponding metric space $\mathcal{W}$ contains isometric copies of $\mathcal{X}$ and $\mathcal{Y}$.

5.12. Observation. *Let $\mathcal{X}$, $\mathcal{Y}$ be metric spaces. Given an appropriate function $f\colon \mathcal{X} \times \mathcal{Y} \to \mathbb{R}$, set*

$$\begin{aligned} a_f &= \max_{x\in\mathcal{X}}\{\min_{y\in\mathcal{Y}}\{f(x,y)\}\}, \\ b_f &= \max_{y\in\mathcal{Y}}\{\min_{x\in\mathcal{X}}\{f(x,y)\}\}, \\ c_f &= \max\{a_f, b_f\}. \end{aligned}$$

Then

$$|\mathcal{X} - \mathcal{Y}|_{\text{GH}} = \inf\{c_f\},$$

where the greatest lower bound is taken for all appropriate functions $f\colon \mathcal{X} \times \mathcal{Y} \to \mathbb{R}$.

Proof of 5.11 Equip the product $\mathcal{X} \times \mathcal{Y}$ with ℓ_1-metric; that is,

$$|(x,y) - (x',y')|_{\mathcal{X}\times\mathcal{Y}} := |x-x'|_{\mathcal{X}} + |y-y'|_{\mathcal{Y}}$$

Note that any appropriate functions $f\colon \mathcal{X} \times \mathcal{Y} \to \mathbb{R}$ is 1-Lipschitz.

Let us equip the space of appropriate functions $\mathcal{X} \times \mathcal{Y} \to \mathbb{R}$ with sup-norm. Observe that the functional $f \mapsto c_f$ is continuous. By the Arzelà–Ascoli theorem, we can choose an appropriate function f with minimal possible value c_f. It remains to apply 5.12. □

5.13. Exercise. Construct three compact metric spaces $\mathcal{X}$, $\mathcal{Y}$, and $\mathcal{Z}$ such that for any metric space $\mathcal{W}$ with subsets $\mathcal{X}'$, $\mathcal{Y}'$, and $\mathcal{Z}'$ such that $\mathcal{X}' \overset{iso}{=\!=} \mathcal{X}$, $\mathcal{Y}' \overset{iso}{=\!=} \mathcal{Y}$, and $\mathcal{Z}' \overset{iso}{=\!=} \mathcal{Z}$ at least one of the following three inequalities is strict

$$\begin{aligned} |\mathcal{X}' - \mathcal{Y}'|_{\text{Haus}\,\mathcal{W}} &\geqslant |\mathcal{X} - \mathcal{Y}|_{\text{GH}}, \\ |\mathcal{Y}' - \mathcal{Z}'|_{\text{Haus}\,\mathcal{W}} &\geqslant |\mathcal{Y} - \mathcal{Z}|_{\text{GH}}, \\ |\mathcal{Z}' - \mathcal{X}'|_{\text{Haus}\,\mathcal{W}} &\geqslant |\mathcal{Z} - \mathcal{X}|_{\text{GH}}. \end{aligned}$$

D Convergence

The Gromov–Hausdorff metric defines Gromov–Hausdorff convergence. Namely, a sequence of compact metric spaces $\mathcal{X}_n$ converges to compact metric spaces $\mathcal{X}_\infty$ in the sense of Gromov–Hausdorff if

$$|\mathcal{X}_n - \mathcal{X}_\infty|_{\mathrm{GH}} \to 0 \quad \text{as} \quad n \to \infty.$$

This convergence is more important than the metric— in all applications, we use only the topology on GH, and we do not care about the particular value of the Gromov–Hausdorff distance between spaces. The following observation follows from 5.10:

5.14. Observation. *A sequence of compact metric spaces $(\mathcal{X}_n)$ converges to $\mathcal{X}_\infty$ in the sense of Gromov–Hausdorff if and only if there is a sequence $\varepsilon_n \to 0+$ and an ε_n-isometry $f_n\colon \mathcal{X}_n \to \mathcal{X}_\infty$ for each n.*

5.15. Exercise.

(a) Show that a circle is not a Gromov–Hausdorff limit of compact simply-connected length spaces.
(b) Construct a compact non-simply-connected metric space that is a Gromov–Hausdorff limit of compact simply-connected length spaces:

5.16. Exercise.

(a) Show that a sequence of length metrics on the 2-sphere cannot converge to the unit disk in the sense of Gromov–Hausdorff.
(b) Construct a sequence of length metrics on the 3-sphere that converges to the unit 3-ball in the sense of Gromov–Hausdorff.

E Uniformly totally bonded families

5.17. Definition. A family $\boldsymbol{Q}$ (isometry classes) of compact metric spaces is called uniformly totally bonded if it meets the following two conditions:

(a) Spaces in $\boldsymbol{Q}$ have uniformly bounded diameters; that is, there is $D \in \mathbb{R}$ such that

$$\operatorname{diam} \mathcal{X} \leqslant D$$

for any space $\mathcal{X}$ in $\boldsymbol{Q}$.
(b) For any $\varepsilon > 0$, there is $n \in \mathbb{N}$ such that any space $\mathcal{X}$ in $\boldsymbol{Q}$ admits an ε-net with at most n points.

5.18. Exercise. Let $\boldsymbol{Q}$ be a family of compact spaces with uniformly bounded diameters. Show that $\boldsymbol{Q}$ is uniformly totally bonded if for any $\varepsilon > 0$ there is $n \in \mathbb{N}$ such that

$$\operatorname{pack}_\varepsilon \mathcal{X} \leqslant n$$

for any space $\mathcal{X}$ in $\boldsymbol{Q}$.

Fix a real constant C. A Borel measure μ on a metric space $\mathcal{X}$ is called C-doubling if

$$\mu[\mathrm{B}(p, 2 \cdot r)] < C \cdot \mu[\mathrm{B}(p, r)]$$

for any point $p \in \mathcal{X}$ and any $r > 0$. A Borel measure is called doubling if it is C-doubling for some real constant C.

5.19. Exercise. Let $\boldsymbol{Q}(C, D)$ be the set of all the compact metric spaces with diameter at most D that admit a C-doubling measure. Show that $\boldsymbol{Q}(C, D)$ is uniformly totally bounded.

Fix an integer constant $M \geqslant 0$. A metric space $\mathcal{X}$ is called M-doubling if any $2 \cdot r$-ball in $\mathcal{X}$ can be covered by M r-balls. A space $\mathcal{X}$ is called doubling if it is M-doubling for some M.

Observe that *a space is doubling if it admits a doubling measure.*

5.20. Exercise. Given a metric space $\mathcal{X}$, consider family $\boldsymbol{B}_{\mathcal{X}}$ of all rescaled balls $\frac{1}{r} \cdot \overline{\mathrm{B}}[x, r]_{\mathcal{X}}$ for all $r > 0$ and $x \in \mathcal{X}$. Show that $\mathcal{X}$ is doubling if and only if $\boldsymbol{B}_{\mathcal{X}}$ is uniformly totally bounded.

Given two metric spaces $\mathcal{X}$ and $\mathcal{Y}$, we will write $\mathcal{X} \leqslant \mathcal{Y}$ if there is a distance-noncontracting map $f: \mathcal{X} \to \mathcal{Y}$; that is, if

$$|x - x'|_{\mathcal{X}} \leqslant |f(x) - f(x')|_{\mathcal{Y}}$$

for any $x, x' \in \mathcal{X}$.

5.21. Exercise.

(a) Let $\mathcal{Y}$ be a compact metric space. Show that the set of all spaces $\mathcal{X}$ such that $\mathcal{X} \leqslant \mathcal{Y}$ is uniformly totally bounded.
(b) Show that for any uniformly totally bounded set $\boldsymbol{Q} \subset \mathrm{GH}$, there is a compact space $\mathcal{Y}$ such that $\mathcal{X} \leqslant \mathcal{Y}$ for any $\mathcal{X}$ in $\boldsymbol{Q}$.

F Gromov selection theorem

The following theorem is analogous to Blaschke selection theorems (4.8).

5.22. Gromov selection theorem. *Let $\boldsymbol{Q}$ be a closed subset of* GH*. Then $\boldsymbol{Q}$ is compact if and only if the spaces in $\boldsymbol{Q}$ are uniformly totally bounded.*

5.23. Lemma. *The space* GH *is complete.*

Suppose $\mathcal{U}$ and $\mathcal{V}$ are metric spaces with isometric closed sets $A \subset \mathcal{U}$ and $A' \subset \mathcal{V}$; let $\iota\colon A \to A'$ be an isometry. Consider the gluing $\mathcal{W} = \mathcal{U} \sqcup_\iota \mathcal{V}$ of $\mathcal{U}$ and $\mathcal{V}$ along ι [see Sect. D].

Let us identify points of $\mathcal{U}$ and $\mathcal{V}$ with their images in $\mathcal{W}$. It is straightforward to check that the metric on $\mathcal{W}$ is defined by

$$\begin{aligned}
|u - u'|_{\mathcal{W}} &:= |u - u'|_{\mathcal{U}},\\
|v - v'|_{\mathcal{W}} &:= |v - v'|_{\mathcal{V}},\\
|u - v|_{\mathcal{W}} &:= \min\left\{\, |u - a|_{\mathcal{U}} + |v - \iota(a)|_{\mathcal{V}} \;:\; a \in A \,\right\},
\end{aligned}$$

where $u, u' \in \mathcal{U}$ and $v, v' \in \mathcal{V}$.

If one applies this construction to two copies of one space $\mathcal{U}$ with a set $A \subset \mathcal{U}$ and the identity map $\iota\colon A \to A$, then the obtained space is called the doubling of $\mathcal{U}$ along A; this space can be denoted by $\sqcup^2_A \mathcal{U}$.

Note that the inclusions $\mathcal{U} \hookrightarrow \mathcal{W}$ and $\mathcal{V} \hookrightarrow \mathcal{W}$ are distance-preserving. Therefore, we can and will consider $\mathcal{U}$ and $\mathcal{V}$ as the subspaces of $\mathcal{W}$; this way the subsets A and A' will be identified and denoted further by A. Note that $A = \mathcal{U} \cap \mathcal{V} \subset \mathcal{W}$.

Proof Let $\mathcal{X}_1, \mathcal{X}_2, \dots$ be a Cauchy sequence in GH. Passing to a subsequence if necessary, we can assume that $|\mathcal{X}_n - \mathcal{X}_{n+1}|_{\mathrm{GH}} < \frac{1}{2^n}$ for each n. In particular, for each n, there is a metric space $\mathcal{V}_n$ with distance-preserving inclusions $\mathcal{X}_n \hookrightarrow \mathcal{V}_n$ and $\mathcal{X}_{n+1} \hookrightarrow \mathcal{V}_n$ such that

$$|\mathcal{X}_n - \mathcal{X}_{n+1}|_{\mathrm{Haus}\,\mathcal{V}_n} < \tfrac{1}{2^n}$$

for each n. Moreover, we may assume that $\mathcal{V}_n = \mathcal{X}_n \cup \mathcal{X}_{n+1}$.

Let us glue $\mathcal{V}_1$ to $\mathcal{V}_2$ along $\mathcal{X}_2$; to the obtained space glue $\mathcal{V}_3$ along $\mathcal{X}_3$, and so on. The obtained metric space $\mathcal{W}$ has an underlying set formed by the disjoint union of all $\mathcal{X}_n$ such that each inclusion $\mathcal{X}_n \hookrightarrow \mathcal{W}$ is distance-preserving and

$$|\mathcal{X}_n - \mathcal{X}_{n+1}|_{\mathrm{Haus}\,\mathcal{W}} < \tfrac{1}{2^n}$$

for each n. In particular,

$$|\mathcal{X}_m - \mathcal{X}_n|_{\mathrm{Haus}\,\mathcal{W}} < \tfrac{1}{2^{n-1}} \tag{2}$$

if $m > n$.

Denote by $\bar{\mathcal{W}}$ the completion of $\mathcal{W}$. Observe that the union $\mathcal{X}_1 \cup \mathcal{X}_2 \cup \cdots \cup \mathcal{X}_n$ is compact and ② implies that it forms a $\frac{1}{2^{n-1}}$-net in $\bar{\mathcal{W}}$. Whence $\bar{\mathcal{W}}$ is compact; see 1.11(c) and 1.13.

Applying the Blaschke selection theorem (4.8), we can pass to a subsequence of $\mathcal{X}_n$ that converges in Haus $\bar{\mathcal{W}}$; denote its limit by $\mathcal{X}_\infty$. It remains to observe that $\mathcal{X}_\infty$ is the Gromov–Hausdorff limit of $\mathcal{X}_n$. □

Proof of 5.22; only-if part Suppose that there is no sequence $\varepsilon_n \to 0$ as described in 5.17. Observe that in this case there is a sequence of spaces $\mathcal{X}_n \in \boldsymbol{Q}$ such that

$$\operatorname{pack}_\delta \mathcal{X}_n \to \infty \quad \text{as} \quad n \to \infty$$

for some fixed $\delta > 0$.

Since $\boldsymbol{Q}$ is compact, this sequence has a partial limit, say $\mathcal{X}_\infty \in \boldsymbol{Q}$. Observe that $\operatorname{pack}_\delta \mathcal{X}_\infty = \infty$. Therefore, $\mathcal{X}_\infty$ is not compact—a contradiction.

If Part Given a positive integer n, consider the set of all nonempty metric spaces $\mathcal{W}_n$ with the number of points at most n and diameter $\leqslant D$. Note that $\mathcal{W}_n$ is a compact set in GH for each n.

Let D and $n = n(\varepsilon)$ be as in the definition of uniformly totally bonded families (5.17).

Note that an ε-net of any $\mathcal{X} \in \boldsymbol{Q}$ belongs to $\mathcal{W}_{n(\varepsilon)}$. Therefore, $\mathcal{W}_{n(\varepsilon)}$ is a compact ε-net of $\boldsymbol{Q}$ for any $\varepsilon > 0$. Since $\boldsymbol{Q}$ is closed in a complete space GH, it implies that $\boldsymbol{Q}$ is compact. □

5.24. Exercise. Show that most of the compact metric spaces are homeomorphic to the Cantor set.

More precisely, suppose $\boldsymbol{Q}$ denotes all metric spaces homeomorphic to the Cantor set. Show that $\boldsymbol{Q}$ is a dense G-delta set in GH.

5.25. Exercise. Show that the space GH is:

(a) Separable (b) Length, and (c) Geodesic

5.26. Exercise. For two metric spaces $\mathcal{X}$ and $\mathcal{Y}$, we write $\mathcal{X} \leqslant \mathcal{Y} + \varepsilon$ if there is a map $f : \mathcal{X} \to \mathcal{Y}$ such that

$$|x - x'|_{\mathcal{X}} \leqslant |f(x) - f(x')|_{\mathcal{Y}} + \varepsilon$$

for any $x, x' \in \mathcal{X}$:

(a) Show that

$$|\mathcal{X} - \mathcal{Y}|_{\mathrm{GH}'} := \inf\{\, \varepsilon > 0 \;:\; \mathcal{X} \leqslant \mathcal{Y} + \varepsilon \quad \text{and} \quad \mathcal{Y} \leqslant \mathcal{X} + \varepsilon \,\}$$

defines a metric on the space of (isometry classes) compact metric spaces.

(b) Moreover, $|{*}-{*}|_{\mathrm{GH}'}$ is equivalent to the Gromov–Hausdorff metric; that is,

$$|\mathcal{X}_n - \mathcal{X}_\infty|_{\mathrm{GH}} \to 0 \quad \iff \quad |\mathcal{X}_n - \mathcal{X}_\infty|_{\mathrm{GH}'} \to 0$$

as $n \to \infty$.

G Universal ambient space

Recall that a metric space is called universal if it contains an isometric copy of any separable metric space (in particular, any compact metric space). Examples of universal spaces include $\mathcal{U}_\infty$—the Urysohn space and ℓ^∞—the space of bounded infinite sequences with the metric defined by sup-norm; see 2.12 and 2.3.

The following proposition says that the space $\mathcal{W}$ in Definition 5.1 can be exchanged to a fixed universal space.

5.27. Proposition. *Let $\mathcal{U}$ be a universal space. Then for any compact metric spaces $\mathcal{X}$ and $\mathcal{Y}$, we have*

$$|\mathcal{X} - \mathcal{Y}|_{\mathrm{GH}} = \inf\{|\mathcal{X}' - \mathcal{Y}'|_{\mathrm{Haus}\,\mathcal{U}}\},$$

where the greatest lower bound is taken over all pairs of sets $\mathcal{X}'$ and $\mathcal{Y}'$ in $\mathcal{U}$ that are isometric to $\mathcal{X}$ and $\mathcal{Y}$, respectively.

Proof of 5.27 By the definition (5.1), we have that

$$|\mathcal{X} - \mathcal{Y}|_{\mathrm{GH}} \leqslant \inf\{|\mathcal{X}' - \mathcal{Y}'|_{\mathrm{Haus}\,\mathcal{U}}\};$$

it remains to prove the opposite inequality.

Suppose $|\mathcal{X} - \mathcal{Y}|_{\mathrm{GH}} < \varepsilon$; let $\mathcal{X}'$, $\mathcal{Y}'$, and $\mathcal{W}$ be as in 5.1. We can assume that $\mathcal{W} = \mathcal{X}' \cup \mathcal{Y}'$; otherwise, pass to the subspace $\mathcal{X}' \cup \mathcal{Y}'$ of $\mathcal{W}$. In this case, $\mathcal{W}$ is compact; in particular, it is separable.

Since $\mathcal{U}$ is universal, there is a distance-preserving embedding of $\mathcal{W}$ in $\mathcal{U}$; let us keep the same notation for $\mathcal{X}'$, $\mathcal{Y}'$, and their images. It follows that

$$|\mathcal{X}' - \mathcal{Y}'|_{\mathrm{Haus}\,\mathcal{U}} < \varepsilon,$$

—hence, the result. □

5.28. Exercise. Let $\mathcal{U}_\infty$ be the Urysohn space. Given two compact sets A and B in $\mathcal{U}_\infty$, define

$$\|A - B\| := \inf\{|A - \iota(B)|_{\mathrm{Haus}\,\mathcal{U}_\infty}\},$$

where the greatest lower bound is taken for all isometrics ι of $\mathcal{U}_\infty$. Show that $\| * - * \|$ defines a semimetric on nonempty compact subsets of $\mathcal{U}_\infty$ and its corresponding metric space is isometric to GH.

The value $\|A - B\|$ is called Hausdorff distance up to isometry from A to B in $\mathcal{U}_\infty$.

H Remarks

Suppose $\mathcal{X}_n \xrightarrow{\text{GH}} \mathcal{X}_\infty$, then there is a metric on the disjoint union

$$\boldsymbol{X} = \bigsqcup_{n \in \mathbb{N} \cup \{\infty\}} \mathcal{X}_n$$

that satisfies the following property:

5.29. Property. *The restriction of metric on each $\mathcal{X}_n$ and $\mathcal{X}_\infty$ coincides with its original metric, and $\mathcal{X}_n \xrightarrow{\text{H}} \mathcal{X}_\infty$ as subsets in $\boldsymbol{X}$.*

Indeed, since $\mathcal{X}_n \xrightarrow{\text{GH}} \mathcal{X}_\infty$, there is a metric on $\mathcal{V}_n = \mathcal{X}_n \sqcup \mathcal{X}_\infty$ such that the restriction of metric on each $\mathcal{X}_n$ and $\mathcal{X}_\infty$ coincides with its original metric, and $|\mathcal{X}_n - \mathcal{X}_\infty|_{\text{Haus } \mathcal{V}_n} < \varepsilon_n$ for some sequence $\varepsilon_n \to 0$. Gluing all $\mathcal{V}_n$ along $\mathcal{X}_\infty$, we get the required space $\boldsymbol{X}$.

In other words, the metric on $\boldsymbol{X}$ *defines* the convergence $\mathcal{X}_n \xrightarrow{\text{GH}} \mathcal{X}_\infty$. This metric makes it possible to talk about limits of sequences $x_n \in \mathcal{X}_n$ as $n \to \infty$, as well as weak limits of a sequence of Borel measures μ_n on $\mathcal{X}_n$ and so on.

For that reason, it is useful to define convergence by specifying the metric on $\boldsymbol{X}$ that satisfies the property for the variation of Hausdorff convergence described in Sect. D.

This approach is more flexible; in particular, it can be used to define the Gromov–Hausdorff convergence of arbitrary metric spaces (not necessarily compact). A limit space for this generalized convergence is not uniquely defined. For example, if each space $\mathcal{X}_n$ in the sequence is isometric to the half-line, then its limit might be isometric to the half-line or the whole line. The first convergence is evident, and the second could be guessed from the diagram.

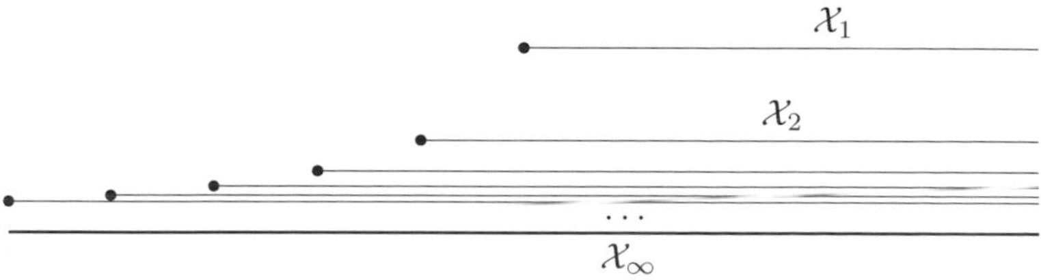

Often the isometry class of the limit can be fixed by marking a point p_n in each space $\mathcal{X}_n$, it is called pointed Gromov–Hausdorff convergence—we say that $(\mathcal{X}_n, p_n)$ converges to $(\mathcal{X}_\infty, p_\infty)$ if there is a metric on $\boldsymbol{X}$ as in 5.29 such that $\mathcal{X}_n \xrightarrow{\mathrm{H}} \mathcal{X}_\infty$ and $p_n \to p_\infty$. For example, the sequence $(\mathcal{X}_n, p_n) = (\mathbb{R}_+, 0)$ converges to $(\mathbb{R}_+, 0)$, while $(\mathcal{X}_n, p_n) = (\mathbb{R}_+, n)$ converges to $(\mathbb{R}, 0)$.

The pointed convergence works nicely for proper metric spaces; the following theorem is an analog of Gromov's selection theorem for this convergence.

5.30. Theorem. *Let $\boldsymbol{Q}$ be a set of isometry classes of pointed proper metric spaces. Assume that for any $R > 0$, the R-balls in the spaces centered at the marked points form a uniformly totally bounded family of spaces. Then $\boldsymbol{Q}$ is precompact with respect to the pointed Gromov–Hausdorff convergence.*

Let us mention a characterization of doubling spaces discovered by Patrice Assouad [4], [25, 12.2].

Given a metric space $\mathcal{X}$, denote by $\mathcal{X}^\theta$ a space with the same underlying set and the metric defined by

$$|x - y|_{\mathcal{X}^\theta} := |x - y|^\theta_{\mathcal{X}};$$

here we assume that $0 < \theta < 1$. The space $\mathcal{X}^\theta$ is called snowflake of $\mathcal{X}$.

5.31. Theorem. *Suppose $0 < \theta < 1$. A metric space $\mathcal{X}$ is doubling if and only if its snowflake $\mathcal{X}^\theta$ admits a bi-Lipschitz embedding in some Euclidean space.*

Lecture 6
Ultralimits

Ultralimits provide a very general way to pass to a limit. This procedure works for *any* sequence of metric spaces, its result reminds limit in the sense of Gromov–Hausdorff, but has some strange features; for example, the limit of a constant sequence of spaces $\mathcal{X}_n = \mathcal{X}$ is *not* $\mathcal{X}$ in general (see 6.13(b)).

In geometry, ultralimits are used mostly as a canonical way to pass to a convergent subsequence. It is very useful in the proofs where one needs to repeat "pass to convergent subsequence" too many times.

This lecture is based on the introductory part of the paper by Bruce Kleiner and Bernhard Leeb [35].

A Faces of ultrafilters

Measure-theoretic definition Recall that $\mathbb{N} = \{1, 2, \dots\}$ is the set of natural numbers.

6.1. Definition. A finitely additive measure ω on $\mathbb{N}$ is called an ultrafilter if it meets the following condition:

(a) $\omega(\mathbb{N}) = 1$ and $\omega(S) = 0$ or 1 for any subset $S \subset \mathbb{N}$.

An ultrafilter ω is called nonprincipal if in addition:

(b) $\omega(F) = 0$ for any finite subset $F \subset \mathbb{N}$.

If $\omega(S) = 0$ for some subset $S \subset \mathbb{N}$, we say that S is ω-small. If $\omega(S) = 1$, we say that S contains ω-almost all elements of $\mathbb{N}$.

6.2. Advanced exercise. Let ω be an ultrafilter on $\mathbb{N}$ and $f : \mathbb{N} \to \mathbb{N}$. Suppose that $\omega(S) \leqslant \omega(f^{-1}(S))$ for any set $S \subset \mathbb{N}$. Show that $f(n) = n$ for ω-almost all $n \in \mathbb{N}$.

A. Petrunin, *Pure Metric Geometry*, SpringerBriefs in Mathematics,
https://doi.org/10.1007/978-3-031-39162-0_6

Classical definition More commonly, a nonprincipal ultrafilter is defined as a collection, say $\mathfrak{F}$, of subsets in $\mathbb{N}$ such that:

1. If $P \in \mathfrak{F}$ and $Q \supset P$, then $Q \in \mathfrak{F}$.
2. If $P, Q \in \mathfrak{F}$, then $P \cap Q \in \mathfrak{F}$.
3. For any subset $P \subset \mathbb{N}$, either P or its complement is an element of $\mathfrak{F}$.
4. If $F \subset \mathbb{N}$ is finite, then $F \not\in \mathfrak{F}$.

Setting $P \in \mathfrak{F} \Leftrightarrow \omega(P) = 1$ makes these two definitions equivalent.

A nonempty collection of sets $\mathfrak{F}$ that does not include the empty set and satisfies only conditions 1 and 2 is called a filter; if, in addition, $\mathfrak{F}$ satisfies condition 3, it is called an ultrafilter. From Zorn's lemma, it follows that every filter contains an ultrafilter. Thus there is an ultrafilter $\mathfrak{F}$ contained in the filter of all complements of finite sets. Clearly, this ultrafilter $\mathfrak{F}$ is nonprincipal.

Stone–Čech compactification Given a set $S \subset \mathbb{N}$, consider subset Ω_S of all ultrafilters ω such that $\omega(S) = 1$. It is straightforward to check that the sets Ω_S for all subsets $S \subset \mathbb{N}$ form a topology on the set of ultrafilters on $\mathbb{N}$. The obtained space was first considered by Andrey Tikhonov and called Stone–Čech compactification of $\mathbb{N}$; it is usually denoted as $\beta\mathbb{N}$.

Let ω_n be a principal ultrafilter such that $\omega_n(\{n\}) = 1$; that is, $\omega_n(S) = 1$ if and only if $n \in S$. Note that $n \mapsto \omega_n$ defines an embedding $\mathbb{N} \hookrightarrow \beta\mathbb{N}$; so, we can (and will) consider $\mathbb{N}$ as a subset of $\beta\mathbb{N}$.

The space $\beta\mathbb{N}$ is the maximal compact Hausdorff space that contains $\mathbb{N}$ as an everywhere dense subset. More precisely, the inclusion $\mathbb{N} \hookrightarrow \beta\mathbb{N}$ has the following universal property: *for any compact Hausdorff space $\mathcal{X}$ and a map $f : \mathbb{N} \to \mathcal{X}$ there is a unique continuous map $\bar{f} : \beta\mathbb{N} \to X$ such that the restriction $\bar{f}|_{\mathbb{N}}$ coincides with f.*

B Ultralimits of points

Let us fix a nonprincipal ultrafilter ω once and for all.

Assume x_n is a sequence of points in a metric space $\mathcal{X}$. Let us define the ω-limit of a sequence $x_1, x_2, \dots$ as the point $x_\omega \in \mathcal{X}$ such that for any $\varepsilon > 0$, point x_n lies in $\mathrm{B}(x_\omega, \varepsilon)$ for ω-almost all n; that is, if

$$S_\varepsilon = \left\{ n \in \mathbb{N} : |x_\omega - x_n| < \varepsilon \right\},$$

then $\omega(S_\varepsilon) = 1$ for any $\varepsilon > 0$. In this case, we will write

$$x_\omega = \lim_{n \to \omega} x_n \text{ or } x_n \to x_\omega \text{ as } n \to \omega.$$

For example, if ω_n is the *principal* ultrafilter defined by $\omega_n\{n\} = 1$ for some $n \in \mathbb{N}$, then $x_{\omega_n} = x_n$.

The sequence x_n can be regarded as a map $\mathbb{N} \to \mathcal{X}$ defined by $n \mapsto x_n$. If $\mathcal{X}$ is compact, then the map $\mathbb{N} \to \mathcal{X}$ can be extended to a continuous map $\beta\mathbb{N} \to \mathcal{X}$ from the Stone–Čech compactification $\beta\mathbb{N}$ of $\mathbb{N}$. Then the ω-limit x_ω is the image of ω.

Note that the ω-limits of a sequence and its subsequence may differ. For example, sequence $y_n = -(-1)^n$ is a subsequence of $x_n = (-1)^n$, but for any ultrafilter ω, we have

$$\lim_{n\to\omega} x_n \neq \lim_{n\to\omega} y_n.$$

6.3. Proposition. *Let x_n be a sequence of points in a metric space $\mathcal{X}$. Assume $x_n \to x_\omega$ as $n \to \omega$. Then x_ω is a* partial limit *of x_n; that is, there is a subsequence of x_n that converges to x_ω in the usual sense.*

Proof Given $\varepsilon > 0$, let $S_\varepsilon = \left\{ n \in \mathbb{N} : |x_n - x_\omega| < \varepsilon \right\}$. Recall that $\omega(S_\varepsilon) = 1$ for any $\varepsilon > 0$.

Since ω is nonprincipal, the set S_ε is infinite for any $\varepsilon > 0$. Therefore, we can choose an increasing sequence n_k such that $n_k \in S_{\frac{1}{k}}$ for each $k \in \mathbb{N}$. Clearly, $x_{n_k} \to x_\omega$ as $k \to \infty$. □

6.4. Proposition. *Any sequence x_n of points in a compact metric space $\mathcal{X}$ has a unique ω-limit x_ω.*

In particular, a bounded sequence of real numbers has a unique ω-limit.

The proposition is analogous to the Bolzano–Weierstrass theorem, and it can be proved the same way. The following lemma is an ultralimit analog of the Cauchy convergence test.

6.5. Lemma. *A sequence of points in a metric space converges if and only if all its subsequences have the same ω-limit.*

Proof The only-if part is evident; it remains to prove the if part. Suppose z is a ω-limit of all subsequences of $x_1, x_2, \dots$. By 6.3, z is a partial limit of x_n. If $x_1, x_2, \dots$ is Cauchy, then $x_n \to z$, and the lemma is proved.

Assume $x_1, x_2, \dots$ is not Cauchy. Then for some $\varepsilon > 0$, there is a subsequence y_n of x_n such that $|x_n - y_n| \geqslant \varepsilon$ for all n. Therefore, $|x_\omega - y_\omega| \geqslant \varepsilon$—a contradiction. □

Recall that ℓ^∞ denotes the space of bounded sequences of real numbers equipped with the sup-norm.

6.6. Exercise. Construct a linear functional $L\colon \ell^\infty \to \mathbb{R}$ such that for any sequence $\boldsymbol{s} = (s_1, s_2, \dots) \in \ell^\infty$ the image $L(\boldsymbol{s})$ is a partial limit of $s_1, s_2, \dots$.

6.7. Exercise. Suppose that $f : \mathbb{N} \to \mathbb{N}$ is a map such that

$$\lim_{n\to\omega} x_n = \lim_{n\to\omega} x_{f(n)}$$

for any bounded sequence x_n of real numbers. Show that $f(n) = n$ for ω-almost all $n \in \mathbb{N}$.

C An illustration

In this section, we illustrate the power of ultralimits by proving the following simple claim.

6.8. Claim. Let $\mathcal{X}$ and $\mathcal{Y}$ be compact spaces. Suppose that for every $n \in \mathbb{N}$ there is a $\frac{1}{n}$-isometry $f_n : \mathcal{X} \to \mathcal{Y}$. Then there is an isometry $\mathcal{X} \to \mathcal{Y}$.

Proof Consider the ω-limit f_ω of f_n; according to 6.4, f_ω is defined. Since

$$|f_n(x) - f_n(x')| \lessgtr |x - x'| \pm \tfrac{1}{n}$$

we get that

$$|f_\omega(x) - f_\omega(x')| = |x - x'|$$

for any $x, x' \in \mathcal{X}$; that is, f_ω is distance-preserving.

Further, since f_n is a $\frac{1}{n}$-isometry, for any $y \in \mathcal{Y}$, there is a sequence $x_1, x_2, \dots \in \mathcal{X}$ such that $|f_n(x_n) - y| \leqslant \frac{1}{n}$ for any n. Therefore,

$$f_\omega(x_\omega) = y,$$

where x_ω is the ω-limit of x_n; that is, f_ω is onto.

It follows that $f_\omega : \mathcal{X} \to \mathcal{Y}$ is an isometry. □

D Ultralimits of spaces

Recall that ω is a fixed nonprincipal ultrafilter on $\mathbb{N}$.

Let $\mathcal{X}_n$ be a sequence of metric spaces. Consider all sequences of points $x_n \in \mathcal{X}_n$. On the set of all such sequences, define an ∞-semimetric by

$$① \qquad |(x_n) - (y_n)| := \lim_{n\to\omega} |x_n - y_n|_{\mathcal{X}_n}.$$

Note that the ω-limit on the right-hand side is always defined and takes a value in $[0,\infty]$. (The ω-convergence to ∞ is defined analogously to the usual convergence to ∞; that is, $\lim x_n = \infty \iff \lim \frac{1}{x_n} = 0$).

Let $\mathcal{X}_\omega$ be the corresponding metric space; that is, the underlying set of $\mathcal{X}_\omega$ is formed by classes of equivalence of sequences of points $x_n \in \mathcal{X}_n$ defined by

$$(x_n) \sim (y_n) \Leftrightarrow \lim_{n\to\omega} |x_n - y_n| = 0,$$

and the distance is defined by ➀.

The space $\mathcal{X}_\omega$ is called the ω-limit of $\mathcal{X}_n$. (It is also called ω-product; this term is motivated by the fact that $\mathcal{X}_\omega$ is a quotient of the product $\prod \mathcal{X}_n$). Typically, $\mathcal{X}_\omega$ will denote the ω-limit of sequence $\mathcal{X}_n$; we may also write

$$\mathcal{X}_n \to \mathcal{X}_\omega \text{ as } n \to \omega \text{ or } \mathcal{X}_\omega = \lim_{n\to\omega} \mathcal{X}_n.$$

Given a sequence $x_n \in \mathcal{X}_n$, we will denote by x_ω its equivalence class which is a point in $\mathcal{X}_\omega$; it can be written as

$$x_n \to x_\omega \text{ as } n \to \omega, \text{ or } x_\omega = \lim_{n\to\omega} x_n.$$

6.9. Observation. *The ω-limit of any sequence of metric spaces is complete.*

We will repeat the proof of 1.9 using a slightly different language.

Proof Let $\mathcal{X}_n$ be a sequence of metric spaces and $\mathcal{X}_n \to \mathcal{X}_\omega$ as $n \to \omega$. Choose a Cauchy sequence $x_1, x_2, \ldots \in \mathcal{X}_\omega$. Passing to a subsequence, we can assume that $|x_k - x_m|_{\mathcal{X}_\omega} < \frac{1}{k}$ if $k < m$.

Choose a double sequence $x_{n,m} \in \mathcal{X}_n$ such that for any fixed m we have $x_{n,m} \to x_m$ as $n \to \omega$. Note that for any $k < m$ the inequality $|x_{n,k} - x_{n,m}| < \frac{1}{k}$ holds for ω-almost all n.

Given $m \in \mathbb{N}$, consider the subset $S_m \subset \mathbb{N}$ defined by

$$S_m = \left\{ n \geqslant m \ : \ |x_{n,k} - x_{n,l}| < \tfrac{1}{k} \quad \text{for all} \quad k < l \leqslant m \right\}.$$

Note that:

- $\mathbb{N} = S_1 \supset S_2 \supset \ldots$.
- $\omega(S_m) = 1$ for each m.
- $\min S_m \geqslant m$.

Consider the sequence $y_n = x_{n,m(n)}$, where $m(n)$ is the largest value such that $n \in S_{m(n)}$; from above, $m(n) \leqslant n$. Denote by $y_\omega \in \mathcal{X}_\omega$ the ω-limit of y_n.

Observe that $|y_m - x_{n,m}| < \frac{1}{m}$ for ω-almost all n. It follows that $|x_m - y_\omega| \leqslant \frac{1}{m}$ for any m. Therefore, $x_n \to y_\omega$ as $n \to \infty$. That is, any Cauchy sequence in $\mathcal{X}_\omega$ converges. □

6.10. Observation. *The ω-limit of any sequence of length spaces is geodesic.*

Proof If $\mathcal{X}_n$ is a sequence of length spaces, then for any sequence of pairs $x_n, y_n \in X_n$ there is a sequence of $\frac{1}{n}$-midpoints z_n.

Let $x_n \to x_\omega$, $y_n \to y_\omega$, and $z_n \to z_\omega$ as $n \to \omega$. Note that z_ω is a midpoint of x_ω and y_ω in $\mathcal{X}_\omega$.

By 6.9, $\mathcal{X}_\omega$ is complete. Applying Menger's lemma (1.27), we get the statement. □

6.11. Exercise. Show that an ultralimit of metric trees is a metric tree.

6.12. Exercise. Suppose that $\mathcal{X}_\infty$ and $\mathcal{X}_1, \mathcal{X}_2, \dots$ are compact metric spaces. Assume $\mathcal{X}_n \xrightarrow{\text{GH}} \mathcal{X}_\infty$. Show that $\mathcal{X}_\omega \overset{iso}{=\!=} \mathcal{X}_\infty$.

Pointed limit If $\operatorname{diam} \mathcal{X}_n \to \infty$ as $n \to \omega$, then the metric on $\mathcal{X}_\omega$ takes value ∞; so $\mathcal{X}_\omega$ has at least two metric components.

To specify a metric component in $\mathcal{X}_\omega$, we may choose a sequence of marked points p_n in $\mathcal{X}_n$ and pass to the metric component of its ω-limit p_ω in $\mathcal{X}_\omega$. The obtained metric component $\mathcal{Z} = (\mathcal{X}_\omega)_{p_\omega}$ with marked p_ω is called the pointed ω-limit of $(\mathcal{X}_n, p_n)$. Note that $\mathcal{Z}$ is a genuine metric space.

If, in the definition of ultralimit, we consider only sequences $x_n \in \mathcal{X}_n$ with such that $|p_n - x_n|$ is bounded, then arrive at $\mathcal{Z}$.

For proper metric spaces, there is a relation between the pointed ultralimit and the pointed Gromov–Hausdorff limit introduced in Sect. H. Namely, *if $(\mathcal{X}_\infty, p_\infty)$ and $(\mathcal{X}_1, p_1), (\mathcal{X}_2, p_2), \dots$ are proper pointed metric spaces such that $(\mathcal{X}_n, p_n) \xrightarrow{\text{GH}} (\mathcal{X}_\infty, p_\infty)$ as defined in Sect. H, then $(\mathcal{X}_\infty, p_\infty)$ is isometric to the pointed ω-limit of $(\mathcal{X}_n, p_n)$.* The proof is the same as for 6.12.

E Ultrapower

If all the metric spaces in the sequence are identical $\mathcal{X}_n = \mathcal{X}$, its ω-limit $\lim_{n\to\omega} \mathcal{X}_n$ is denoted by $\mathcal{X}^\omega$ and called ω-power of $\mathcal{X}$ (also known as ω-completion).

6.13. Exercise. For any point $x \in \mathcal{X}$, consider the constant sequence $x_n = x$ and set $\iota(x) = \lim_{n\to\omega} x_n \in \mathcal{X}^\omega$:

(a) Show that $\iota \colon \mathcal{X} \to \mathcal{X}^\omega$ is distance-preserving embedding. (So we can and will consider $\mathcal{X}$ as a subset of $\mathcal{X}^\omega$.)
(b) Show that ι is onto if and only if $\mathcal{X}$ is compact.
(c) Show that if $\mathcal{X}$ is proper, then $\iota(\mathcal{X})$ forms a metric component of $\mathcal{X}^\omega$; that is, a subset of $\mathcal{X}^\omega$ that lies at a finite distance from a given point.

If $\mathcal{X}$ is a genuine metric space, then the metric component of $\mathcal{X}$ in $\mathcal{X}^\omega$ is also called the ultrapower of $\mathcal{X}$; if needed, we may call it the small ultrapower, and the whole space $\mathcal{X}^\omega$ could be called the big ultrapower of $\mathcal{X}$. Note that the small ultrapower of genuine metric space is a genuine metric space. Further, according to (c), *proper metric space is isometric to its small ultrapower*.

Note that (b) implies that the inclusion $\mathcal{X} \hookrightarrow \mathcal{X}^\omega$ is not onto if the space $\mathcal{X}$ is not compact. However, the spaces $\mathcal{X}$ and $\mathcal{X}^\omega$ might be isometric; here is an example:

6.14. Exercise. Let $\mathcal{X}$ be an infinite countable set with discrete metric; that is $|x - y|_{\mathcal{X}} = 1$ if $x \neq y$. Show that:

(a) $\mathcal{X}^\omega$ is not isometric to $\mathcal{X}$, but
(b) $\mathcal{X}^\omega$ is isometric to $(\mathcal{X}^\omega)^\omega$.

6.15. Exercise. Given a nonprincipal ultrafilter ω, construct an ultrafilter ω_1 such that

$$\mathcal{X}^{\omega_1} \overset{iso}{=\!=} (\mathcal{X}^\omega)^\omega$$

for any metric space $\mathcal{X}$.

6.16. Observation. *Let $\mathcal{X}$ be a complete metric space. Then $\mathcal{X}^\omega$ is geodesic space if and only if $\mathcal{X}$ is a length space.*

Proof The if part follows from 6.10; it remains to prove the only-if part.

Assume $\mathcal{X}^\omega$ is geodesic. Then any pair of points $x, y \in \mathcal{X}$ has a midpoint $z_\omega \in \mathcal{X}^\omega$. Fix a sequence of points $z_n \in \mathcal{X}$ such that $z_n \to z_\omega$ as $n \to \omega$.

Note that $|x - z_n|_{\mathcal{X}} \to \frac{1}{2} \cdot |x - y|_{\mathcal{X}}$ and $|y - z_n|_{\mathcal{X}} \to \frac{1}{2} \cdot |x - y|_{\mathcal{X}}$ as $n \to \omega$. In particular, for any $\varepsilon > 0$, the point z_n is an ε-midpoint of x and y for ω-almost all n. It remains to apply Menger's lemma (1.27). □

6.17. Exercise. Assume $\mathcal{X}$ is a complete length space and $p, q \in \mathcal{X}$ cannot be joined by a geodesic in $\mathcal{X}$. Show that there is at least a continuum of distinct geodesics between p and q in the ultrapower $\mathcal{X}^\omega$.

6.18. Exercise. Construct a proper metric space $\mathcal{X}$ such that its big ultrapower $\mathcal{X}^\omega$ is not locally compact.

F Tangent and asymptotic spaces

Choose a space $\mathcal{X}$ and a sequence λ_n of positive numbers. Consider the sequence of rescalings $\mathcal{X}_n = \lambda_n \cdot \mathcal{X} = (\mathcal{X}, \lambda_n \cdot |{*} - {*}|_{\mathcal{X}})$.

Choose a point $p \in \mathcal{X}$ and denote by p_n the corresponding point in $\mathcal{X}_n$. Consider the ω-limit $\mathcal{X}_\omega$ of $\mathcal{X}_n$ (one may denote it by $\lambda_\omega \cdot \mathcal{X}$); set p_ω to be the ω-limit of p_n.

If $\lambda_n \to \infty$ as $n \to \omega$, then the metric component of p_ω in $\mathcal{X}_\omega$ is called λ_ω-tangent space at p and denoted by $\mathrm{T}_p^{\lambda_\omega}\mathcal{X}$ (or $\mathrm{T}_p^\omega\mathcal{X}$ if $\lambda_n = n$).

If $\lambda_n \to 0$ as $n \to \omega$, then the metric component of p_ω is called λ_ω-asymptotic space[1] and denoted by $\operatorname{Asym}\mathcal{X}$ or $\operatorname{Asym}^{\lambda_\omega}\mathcal{X}$. Note that the space $\operatorname{Asym}\mathcal{X}$ and its point p_ω do not depend on the choice of $p \in \mathcal{X}$.

The following exercise states that the constructions above depend on the sequence λ_n and a nonprincipal ultrafilter ω.

6.19. Exercise. Construct a metric space $\mathcal{X}$ with a point p such that the tangent space $\mathrm{T}_p^{\lambda_\omega}\mathcal{X}$ (or its asymptotic cone $\operatorname{Asym}^{\lambda_\omega}\mathcal{X}$) depends on the sequence λ_n and/or ultrafilter ω.

For nice spaces, different choices of the sequence of coefficients and ultrafilter may give the same space; some examples are given in the following exercise.

6.20. Exercise. Let $\mathcal{T} = \operatorname{Asym}\mathcal{L}$, where $\mathcal{L}$ is one of the following spaces:

(i) Lobachevsky plane.
(ii) Lobachevsky space.
(iii) 3-regular metric tree; that is, the degree of any vertex. Assume that each edge has unit length.

(a) Show that $\mathcal{T}$ is a complete metric tree.
(b) Show that $\mathcal{T}$ is one-point-homogeneous; that is, given two points $s, t \in \mathcal{T}$ there is an isometry of $\mathcal{T}$ that maps s to t.
(c) Show that $\mathcal{T}$ has continuum degree at any point; that is, for any point $t \in \mathcal{T}$ the set of connected components of the complement $\mathcal{T} \setminus \{t\}$ has cardinality continuum.

6.21. Exercise. Consider the cylinder $\mathbb{S}^1 \times [0, 1]$ with the standard metric. Let $\mathcal{X}$ be the quotient space $\mathbb{S}^1 \times [0, 1]/\mathbb{S}^1 \times \{0\}$; that is,

$$|(u_1, t_1) - (u_2, t_2)|_{\mathcal{X}} := \min\{\, |(u_1, t_1) - (u_2, t_2)|_{\mathbb{S}^1 \times [0,1]}, t_1 + t_2 \,\}.$$

Describe the ultratangent space $\mathrm{T}_o^\omega\mathcal{X}$, where $o \in \mathcal{X}$ is the point that corresponds to $\mathbb{S}^1 \times \{0\}$.

G Remarks

A nonprincipal ultrafilter ω is called selective if for any partition of $\mathbb{N}$ into sets $\{C_\alpha\}_{\alpha \in \mathcal{A}}$ such that $\omega(C_\alpha) = 0$ for each α, there is a set $S \subset \mathbb{N}$ such that $\omega(S) = 1$ and $S \cap C_\alpha$ is a one-point set for each $\alpha \in \mathcal{A}$.

[1] Often it is called an *asymptotic cone* despite that it is not a cone in general; this name is used since in good cases it has a cone structure.

The existence of a selective ultrafilter follows from the continuum hypothesis [59].

If needed, we may assume that the chosen ultrafilter ω is selective. In this case, *the subsequence* $(x_n)_{n\in S}$ *in 6.3 can be chosen so that* $\omega(S) = 1$.

Semisolutions

1.2. Add four triangle inequalities (1.1(d)).
1.4. Consider the function

$$f(x) = \frac{\operatorname{dist}_A x}{\operatorname{dist}_A x + \operatorname{dist}_B x},$$

where $\operatorname{dist}_A x := \inf_{a \in A} |a - x|$. Show that f is continuous and satisfies the needed property.
1.5. Use 1.4 to construct an approximation of the needed function and pass to a limit or find a proof of the Tietze extension theorem.
1.6; (a) Note that if $\mu(A) = \mu(B) = 0$, then $|A - B| = 0$. Therefore, 1.1(b) does not hold for bounded closed subsets. It is straightforward to check the remaining conditions in 1.1 hold true.
(b) Note that the distance from the empty set to the whole plane is infinite; so the value $|A - B|$ might be infinite. It is straightforward to check the remaining conditions in 1.1.

Remark Metrics of the form $|A - B| = \mu(A \triangle B)$ are very special. In particular, they satisfy the so-called hypermetric inequalities; that is, for any sequence of sets $A_1, \ldots, A_n$ and any sequence of integers $b_1, \ldots, b_n$ such that $\sum_i b_i = 1$, we have

$$\sum_{i,j} b_i \cdot b_j \cdot |A_i - A_j| \leqslant 0.$$

Note that for $n = 3$ and $b_1 = b_2 = -b_3 = 1$, we get the usual triangle inequality. For more on the subject, see [16].

1.7. Choose $\delta > 0$ and an increasing linear bijection $\ell\colon [a, b] \to [c, d]$. Show that ℓ has arbitrarily close increasing piecewise-linear bijection $s\colon [a, b] \to [c, d]$ such that derivative at any point is either $< \delta$ or $> \frac{1}{\delta}$.

A. Petrunin, *Pure Metric Geometry*, SpringerBriefs in Mathematics,
https://doi.org/10.1007/978-3-031-39162-0

Start with the identity map $[0,1] \to [0,1]$; iterate the above construction for smaller and smaller δ and pass to the limit. This way we obtain an increasing bijection $x \leftrightarrow x'$ from $[0,1]$ to itself such that for any $\varepsilon > 0$ there is a partition $0 = t_0 < t_1 < \dots < t_{2\cdot n} = 1$ of $[0,1]$ with

$$\begin{aligned}\varepsilon > |t_0 - t_1| + |t_1' - t_2'| + |t_2 - t_3| + \dots \\ \dots + |t_{2\cdot n-2} - t_{2\cdot n-1}| + |t_{2\cdot n-1}' - t_{2\cdot n}'|.\end{aligned}$$

Make a conclusion.

1.8. Assume the statement is wrong. Then for any point $x \in \mathcal{X}$, there is a point $x' \in \mathcal{X}$ such that

$$|x - x'| < \rho(x)$$

and

$$\rho(x') \leqslant \frac{\rho(x)}{1+\varepsilon}.$$

Consider a sequence $x_1, x_2, \dots \in \mathcal{X}$ such that $x_{n+1} = x_n'$. Show that this is a Cauchy sequence. Since $\mathcal{X}$ is complete, x_n converges; denote its limit by x_∞. Since ρ is a continuous function, we get

$$\rho(x_\infty) = \lim_{n\to\infty} \rho(x_n) = 0.$$

The latter contradicts that $\rho > 0$.

1.9. Let $\bar{\mathcal{X}}$ be the completion of $\mathcal{X}$. By the definition, for any $y \in \bar{\mathcal{X}}$, there is a Cauchy sequence x_n in $\mathcal{X}$ that converges to y.

Choose a Cauchy sequence y_m in $\bar{\mathcal{X}}$. From above, we can choose points $x_{n,m} \in \mathcal{X}$ such that $x_{n,m} \to y_m$ for any m. Choose $z_m = x_{n_m,m}$ such that $|y_m - z_m| < \frac{1}{m}$. Observe that z_m is Cauchy. Therefore, its limit z_∞ lie in $\bar{\mathcal{X}}$. Finally, show that $x_m \to z_\infty$.

1.13. A compact ε-net N in $\mathcal{K}$ contains a finite ε-net F. Show and use that F is a $2\cdot\varepsilon$-net of $\mathcal{K}$.

1.15. Given a pair of points $x_0, y_0 \in \mathcal{K}$, consider two sequences $x_0, x_1, \dots$ and $y_0, y_1, \dots$ such that $x_{n+1} = f(x_n)$ and $y_{n+1} = f(y_n)$ for each n.

Since $\mathcal{K}$ is compact, we can choose an increasing sequence of integers n_k such that both sequences $(x_{n_i})_{i=1}^\infty$ and $(y_{n_i})_{i=1}^\infty$ converge. In particular, both are Cauchy; that is,

$$|x_{n_i} - x_{n_j}|_{\mathcal{K}} \to 0 \quad \text{and} \quad |y_{n_i} - y_{n_j}|_{\mathcal{K}} \to 0$$

as $\min\{i,j\} \to \infty$.

Since f is distance-noncontracting,

$$|x_0 - x_{|n_i - n_j|}| \leqslant |x_{n_i} - x_{n_j}|$$

for any i and j. Therefore, there is a sequence $m_i \to \infty$ such that

$$\circledast \qquad x_{m_i} \to x_0 \quad \text{and} \quad y_{m_i} \to y_0$$

as $i \to \infty$.

Since f is distance-noncontracting, the sequence $\ell_n = |x_n - y_n|_{\mathcal{K}}$ is nondecreasing. By $(*)$, $\ell_{m_i} \to \ell_0$ as $m_i \to \infty$. It follows that

$$\ell_0 = \ell_1 = \dots.$$

In particular,

$$|x_0 - y_0|_{\mathcal{K}} = \ell_0 = \ell_1 = |f(x_0) - f(y_0)|_{\mathcal{K}}$$

for any pair of points (x_0, y_0) in $\mathcal{K}$. That is, the map f is distance-preserving; hence, f is injective. From $(*)$, we also get that $f(\mathcal{K})$ is everywhere dense. Since $\mathcal{K}$ is compact, $f : \mathcal{K} \to \mathcal{K}$ is surjective—hence the result.

Remarks This is a basic lemma in the introduction to Gromov–Hausdorff distance [see 7.3.30 in 13]. The presented proof is not quite standard, I learned it from Travis Morrison, a student in my MASS class at Penn State, Fall 2011.

Note that this exercise implies that *any surjective non-expanding map from a compact metric space to itself is an isometry.*

1.16. Check an infinite set with a discrete metric.

1.17. Set $B_p = B(x, \frac{\pi}{2})_{\mathbb{S}^2}$. The triangle inequality follows since

$$\circledast \qquad (B_x \setminus B_y) \cup (B_y \setminus B_z) \supseteq B_x \setminus B_z.$$

The remaining conditions in Definition 1.1 are evident.

Observe that $B_x \setminus B_y$ does not overlap with $B_y \setminus B_z$, and we get equality in $(*)$ if and only if y lies on the great circle arc from x to z. Therefore, the second statement follows.

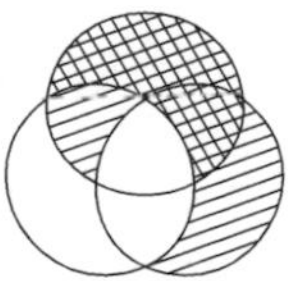

Remarks This construction is due to Aleksei Pogorelov [57]. It is closely related to the construction given by David Hilbert [26] that was the motivating example for his fourth problem. See also the remark after the solution of 1.6.

1.18. We may assume that none of the points p, x, y, z lies on a geodesic between the other two.

Let K be the set in the tree covered by all six geodesics with the endpoints p, x, y, z. Observe that K looks like an H or like an X; make a conclusion.

Remarks The value $\frac{1}{2}\cdot(|p-x|+|p-y|-|x-y|)$ is called Gromov's product of x and y with the origin at p; usually, it is denoted by $(x|y)_p$.

Note that a four-point metric space admits an isometric embedding into a metric tree if and only if one of these two equivalent conditions holds. Moreover, a metric space admits an isometric embedding into a metric tree if every four-point subspace admits such embedding.

1.19. Apply 1.18.

1.22. Note that $\mathcal{P}$ is complete. Choose $\varepsilon > 0$. Use 1.21 to show that the set of paths of length $> 1-\varepsilon$ is open in $\mathcal{P}$; show that this set is also dense in $\mathcal{P}$. Apply Baire's theorem (1.10).

Remark You might find it surprising that *most of the short maps from the sphere to the plane are length-preserving*; that is, they preserve lengths of all curves. The latter follows from the result of Bernd Kirchheim, Emanuele Spadaro, and László Székelyhidi [36]. (While most of the maps have this property, it is not at all easy to construct a single such example.)

1.24. *Formally speaking, a one-point space is a solution, but we will construct a nontrivial example.*

Recall that $c_0 \subset \ell^\infty$ denotes the space of all real sequences converging to zero. Consider the unit ball B in c_0; denote by ρ_0 the metric on B.

Let

$$\varphi(\boldsymbol{x}) = 2 + \tfrac{1}{2}\cdot x_1 + \tfrac{1}{4}\cdot x_2 + \tfrac{1}{8}\cdot x_3 + \dots,$$

where $\boldsymbol{x} = (x_1, x_2 \dots) \in B$. Consider another length metric ρ_1 on B that is different from ρ_0 by the conformal factor φ; that is, if $t \mapsto \boldsymbol{x}(t)$ for $t \in [0,\ell]$ is a curve parametrized by ρ_0-length, then its ρ_1-length is defined by

$$\operatorname{length}_{\rho_1}\boldsymbol{x} := \int_0^\ell \varphi\circ\boldsymbol{x}(t)\cdot dt.$$

Note that the metric ρ_1 is bi-Lipschitz to ρ_0.

Assume $t \mapsto \boldsymbol{x}(t)$ and $t \mapsto \boldsymbol{x}'(t)$ are two curves parametrized by ρ_0-length that differ only in the m-th coordinate; denote them by $x_m(t)$ and $x'_m(t)$, respectively. Show that if $x'_m(t) \leqslant x_m(t)$ for any t and the function $x'_m(t)$ is locally 1-Lipschitz at all t such that $x'_m(t) < x_m(t)$, then

$$\operatorname{length}_{\rho_1}\boldsymbol{x}' \leqslant \operatorname{length}_{\rho_1}\boldsymbol{x}.$$

Moreover, this inequality is strict if $x'_m(t) < x_m(t)$ for some t.

Fix a curve $\boldsymbol{x}(t)$, $t \in [0, \ell]$, parametrized by ρ_0-length. We can choose large m so that $x_m(t)$ is sufficiently close to 0 for any t. In this case, it is easy to construct a function $t \mapsto x'_m$ that meets the above conditions. It follows that for any curve $\boldsymbol{x}(t)$ in (B, ρ_1), we can find a shorter curve $\boldsymbol{x}'(t)$ with the same endpoints. In particular, (B, ρ_1) has no geodesics.

Remark This solution was suggested by Fedor Nazarov [48].

1.25. Choose a sequence of positive numbers $\varepsilon_n \to 0$ and a finite ε_n-net N_n of K for each n. We can assume that $\varepsilon_0 > \operatorname{diam} K$, and N_0 is a one-point set. If $|x - y| < \varepsilon_k$ for some $x \in N_{k+1}$ and $y \subset N_k$, then connect them by a curve of length at most ε_k.

Let K' be the union of all these curves and K. Show that K' is compact and path-connected.

Source This problem is due to Eugene Bilokopytov [6].

1.26. Choose a Cauchy sequence x_n in $(\mathcal{X}, \| * - * \|)$; it is sufficient to show that a subsequence of x_n converges.

Observe that the sequence x_n is Cauchy in $(\mathcal{X}, | * - * |)$; denote its limit by x_∞.

Passing to a subsequence, we can assume that $\|x_n - x_{n+1}\| < \frac{1}{2^n}$. It follows that there is a 1-Lipschitz path γ in $(\mathcal{X}, \| * - * \|)$ such that $x_n = \gamma(\frac{1}{2^n})$ for each n and $x_\infty = \gamma(0)$. Therefore,

$$\|x_\infty - x_n\| \leqslant \operatorname{length} \gamma|_{[0, \frac{1}{2^n}]} \leqslant \tfrac{1}{2^n}.$$

In particular, x_n converges to x_∞ in $(\mathcal{X}, \| * - * \|)$.

Source Thakyin Hu and W.A. Kirk [27, Corollary]; see also [53, Lemma 2.3].

1.30. Let U be the ε-neighborhood of $\mathrm{B}(x, R)_{\mathcal{X}}$. By the triangle inequality, $U \subset \mathrm{B}(x, R + \varepsilon)_{\mathcal{X}}$; this inclusion holds in any metric space.

Choose $y \in \mathrm{B}(x, R+\varepsilon)_{\mathcal{X}}$, so $|x - y|_{\mathcal{X}} < R+\varepsilon$. Since $\mathcal{X}$ is a length space, there is a curve γ from x to y with length less than $R + \varepsilon$. Show and use that γ contains a point m such that $|x - m|_{\mathcal{X}} < R$ and $|y - m|_{\mathcal{X}} < \varepsilon$.

1.31. Consider the following subset of $\mathbb{R}^2$ equipped with the induced length metric

$$\mathcal{X} = \big((0, 1] \times \{0, 1\}\big) \cup \big(\{1, \tfrac{1}{2}, \tfrac{1}{3}, \dots\} \times [0, 1]\big).$$

Note that $\mathcal{X}$ is locally compact and geodesic.

Its completion $\bar{\mathcal{X}}$ is isometric to the closure of $\mathcal{X}$ equipped with the induced length metric. Note that $\bar{\mathcal{X}}$ is obtained from $\mathcal{X}$ by adding two points $p = (0, 0)$ and $q = (0, 1)$.

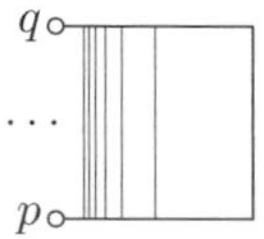

Observe that p admits no compact neighborhood in $\bar{\mathcal{X}}$, and there is no geodesic connecting p to q in $\bar{\mathcal{X}}$.

Source Martin Bridson and André Haefliger [11, I.3.6(4)].

1.32. Suppose this number does not exist. Show that there are two point arrays $(x_1, \dots, x_n)$ and $(y_1, \dots, y_m)$ such that

$$\circledast \qquad \min_{z\in\mathcal{X}}\{\, f(z)\,\} > \max_{z\in\mathcal{X}}\{\, h(z)\,\},$$

where

$$f(z) = \tfrac{1}{n}\cdot\sum_i |x_i - z|_{\mathcal{X}}$$

and

$$h(z) = \tfrac{1}{m}\cdot\sum_j |y_j - z|_{\mathcal{X}}.$$

Note that

$$\begin{aligned}\tfrac{1}{m}\cdot\sum_j f(y_j) &= \tfrac{1}{m\cdot n}\cdot\sum_{i,j}|x_i - y_j|_{\mathcal{X}} = \\ &= \tfrac{1}{n}\cdot\sum_i h(x_i);\end{aligned}$$

that is, the average value of $f(y_j)$ coincides with the average value of $h(x_i)$. The latter contradicts ($*$).

Remark The value ℓ is uniquely defined; it is called the rendezvous value of $\mathcal{X}$. This is a result of Oliver Gross [23].

2.2. By the Fréchet lemma (2.1), we can identify $\mathcal{K}$ with a compact subset in ℓ^∞.

Denote by $\mathcal{L}$ the closed convex hull of $\mathcal{K}$; that is, $\mathcal{L}$ is the minimal convex closed set in ℓ^∞ that contains $\mathcal{K}$. (In other words, $\mathcal{L}$ is the minimal closed set containing $\mathcal{K}$ such that if $x, y \in \mathcal{L}$, then $t\cdot x + (1-t)\cdot y \in \mathcal{L}$ for any $t \in [0, 1]$.)

Observe that $\mathcal{L}$ is a length space. It remains to show that $\mathcal{L}$ is compact.

By construction, $\mathcal{L}$ is a closed subset of ℓ^∞; in particular, it is complete. By 1.11(c), it remains to show that $\mathcal{L}$ is totally bounded.

Recall that Minkowski sum $A+B$ of two sets A and B in a vector space is defined by

$$A + B := \{\, a + b \,:\, a \in A,\ b \in B\,\}.$$

Observe that the Minkowski sum of two convex sets is convex.

Denote by $\bar{B}_\varepsilon$ the closed ε-ball in ℓ^∞ centered at the origin. Choose a finite ε-net N in $\mathcal{K}$ for some $\varepsilon > 0$. Note that $P = \operatorname{Conv} N$ is a convex polyhedron; in particular, $\operatorname{Conv} N$ is compact.

Observe that $N + \bar{B}_\varepsilon$ is a closed ε-neighborhood of N. It follows that $N + \bar{B}_\varepsilon \supset K$, and therefore, $P + \bar{B}_\varepsilon \supset \mathcal{L}$. In particular, P is a $2 \cdot \varepsilon$-net in $\mathcal{L}$; since P is compact and $\varepsilon > 0$ is arbitrary, $\mathcal{L}$ is totally bounded (see 1.13).

Remark Alternatively, one may use that *the injective envelope of a compact space is compact*; see 3.3(b), 3.20, and 3.23.

2.3. Modify the proof of 2.1.

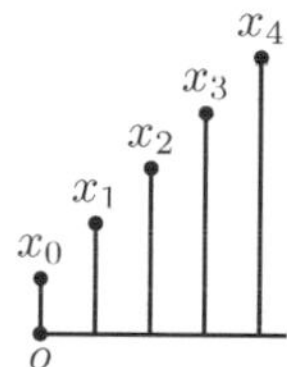

2.8. Consider the metric tree $\mathcal{T}$ shown on the diagram; it is a half-line $[0, \infty)$ with attached an interval of length $n + 1$ to each integer $n \geqslant 0$. Denote by o the origin of the half-line and by x_n the endpoint of nth interval.

Observe that if $m \neq n$, then

$$|x_m - x_n|_{\mathcal{T}} \geqslant |o - x_n|_{\mathcal{T}} + 1.$$

Show and use that for any binary sequence ε_n there is an extension function f such that

$$f(x_n) = |o - x_n|_{\mathcal{T}} + \varepsilon_n.$$

Remark An if-and-only-if condition on $\mathcal{X}$ that have separable $\mathcal{X}^\infty$ was found by Julien Melleray [42, 2.8]. A similar condition was used by Herbert Federer to describe metric spaces where Besicovitch covering lemma holds [18, 2.8.9].

2.13. Choose a separable space $\mathcal{X}$ that has an infinite number of geodesics between a pair of points with the given distance between them; say a square in $\mathbb{R}^2$ with ℓ^∞-metric will do. Apply to $\mathcal{X}$ universality of Urysohn space (2.12).

2.14. First let us prove the following claim:

- Suppose $f : K \to \mathbb{R}$ is an extension function defined on a compact subset K of the Urysohn space $\mathcal{U}$. Then there is a point $p \in \mathcal{U}$ such that $|p - x| = f(x)$ for any $x \in K$.

Without loss of generality, we may assume that $f > 0$. Since K is compact, we may fix $\varepsilon > 0$ such that $f(x) > \varepsilon$ for any $x \in K$.

Consider the sequence $\varepsilon_n = \frac{\varepsilon}{100\cdot 2^n}$. Choose a sequence of ε_n-nets $N_n \subset K$. Applying the universality of $\mathcal{U}$ recursively, we may choose a point p_n such that $|p_n - x| = f(x)$ for any $x \in N_n$ and $|p_n - p_{n-1}| = 10 \cdot \varepsilon_{n-1}$. Observe that the sequence p_n is Cauchy and its limit p meets $|p - x| = f(x)$ for any $x \in K$.

Now, choose a sequence x_n of points that is dense in $\mathcal{S}$. Applying the claim, we may extend the map from K to $K \cup \{x_1\}$, further to $K \cup \{x_1, x_2\}$, and so on. As a result, we extend the distance-preserving map f to the whole sequence x_n. It remains to extend it continuously to the whole space $\mathcal{S}$.

2.16. It is sufficient to show that any compact subspace $\mathcal{K}$ of the Urysohn space $\mathcal{U}$ can be contracted to a point.

Note that any compact space $\mathcal{K}$ can be extended to a contractible compact space $\mathcal{K}'$; for example, we may embed $\mathcal{K}$ into ℓ^∞ and pass to its convex hull, as it was done in 2.2.

By 2.20, there is an isometric embedding of $\mathcal{K}'$ that agrees with the inclusion $\mathcal{K} \hookrightarrow \mathcal{U}$. Since $\mathcal{K}$ is contractible in $\mathcal{K}'$, it is contractible in $\mathcal{U}$.

A Better Way One can contract the whole Urysohn space using the following construction.

Note that points in $\mathcal{X}_\infty$ constructed in the proof of 2.7 can be multiplied by $t \in [0, 1]$—simply multiply each function by t. That defines a map

$$\lambda_t \colon \mathcal{X}_\infty \to \mathcal{X}_\infty$$

that rescales all distances by factor t. The map λ_t can be extended to the completion of $\mathcal{X}_\infty$, which is isometric to $\mathcal{U}_d$ (or $\mathcal{U}$).

Observe that the map λ_1 is the identity and λ_0 maps the whole space to a single point, say x_0—this is the only point of $\mathcal{X}_0$. Further, note that $(t, p) \mapsto \lambda_t(p)$ is a continuous map; in particular, $\mathcal{U}_d$ and $\mathcal{U}$ are contractible.

As a bonus, observe that for any point $p \in \mathcal{U}_d$, the curve $t \mapsto \lambda_t(p)$ is a geodesic path from p to x_0.

Source Mikhael Gromov [22, (d) on page 82].

2.18. Consider two infinite metric trees as on the diagram.

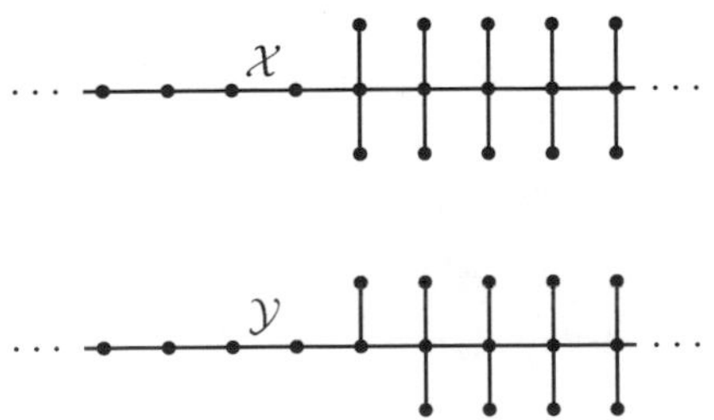

Remark A more sophisticated example: $\mathcal{X} = \ell^\infty$ and $\mathcal{Y} = L^\infty([0, 1])$. Try to prove that it qualifies; see also [12].

2.19; ((a)) and ((b)). Observe that L and M satisfy the definition of d-Urysohn space and apply the uniqueness (2.17). Note that

$$\ell = \operatorname{diam} L = \min\{2 \cdot r, d\}.$$

(c) Use ((a)), maybe twice.

2.21. Let p be the center of the sphere; without loss of generality, we can assume that $|p - x| \leqslant |p - y|$.

Consider function $f\colon \{p, x, y\} \to \mathbb{R}$ defined by $f(p) = 1$, $f(x) = 1 + |p - x|$, and $f(y) = 1 + |p - y| - \varepsilon$. Suppose $\varepsilon > 0$ is sufficiently small; show that f is an extension function on $\{p, x, y\}$.

By the extension property, there is a point $z \in \mathcal{U}$ such that $|p - z| = f(p)$, $|x - z| = f(x)$, and $|y - z| = f(y)$. Whence the statement follows.

Source This problem is taken from a survey of Julien Melleray [42, Prop. 4.3], where it was attributed to Matatyahu Rubin.

2.22. Observe that the complement $\mathcal{V} = \mathcal{U} \setminus B$ is complete. Show that it $\mathcal{V}$ satisfies the extension property. Conclude that $\mathcal{V}$ is an Urysohn space and apply 2.17.

For the second part, observe that there is an isometry $\iota\colon \mathcal{U} \to \mathcal{V}$. Moreover, if p is the center of B, then we can assume that ι has a fixed point x such that $|p - x| > 2$.

Consider the unit sphere S centered at x. The restriction of ι to S is an isometry of S. Use 2.21 to show that it cannot be extended to an isometry of $\mathcal{U}$.

Source Julien Melleray [42, Sec. 4.4].

2.23. Apply 2.17 and the construction in 2.11.

2.24; ((a)). The Euclidean plane is homogeneous in every sense.

((b)) The Hilbert space ℓ^2 is finite-set-homogeneous, but not compact-set-homogeneous, nor countable-set-homogeneous.

((c)) ℓ^∞ is one-point-homogeneous, but not two-point-homogeneous. Try to show that there is no isometry of ℓ^∞ such that

$$(0, 0, 0, \dots) \mapsto (0, 0, 0, \dots),$$
$$(1, 1, 1, \dots) \mapsto (1, 0, 0, \dots).$$

((d)) ℓ^1 is one-point-homogeneous, but not two-point-homogeneous. Try to show that there is no isometry of ℓ^∞ such that

$$(0, 0, 0, \dots) \mapsto (0, 0, 0, \dots),$$
$$(2, 0, 0 \dots) \mapsto (1, 1, 0, \dots).$$

2.25. Let $\mathcal{T}$ be a one-point-homogeneous metric tree. Note that all points in $\mathcal{T}$ have the same degree d; that is, for any point $t \in \mathcal{T}$, the set of connected components of the complement $\mathcal{T} \setminus \{t\}$ has the same cardinality d.

Show that if $d = 0$, then $\mathcal{T}$ is a one-point space; there is no tree with $d = 1$, and if $d = 2$, then $\mathcal{T} \overset{iso}{=\!=} \mathbb{R}$.

Suppose $d \geqslant 3$. Choose a geodesic γ in $\mathcal{T}$. Show that the number of connected components of $\mathcal{T}\backslash\gamma$ has cardinality continuum. Observe and use that one can choose a point p_α in each connected component such that $|p_\alpha - p_\beta|_{\mathcal{T}} > 1$ if $\alpha \neq \beta$.

2.26. Assume $F_{\mathcal{X}}$ is not an embedding. That is, there is a sequence of points $x_1, x_2, \ldots$ and a point x_∞ such that $f_{x_n} \to f_{x_\infty}$ in $C(\mathcal{X}, \mathbb{R})$ as $n \to \infty$, while $|x_n - x_\infty|_{\mathcal{X}} > \varepsilon$ for some fixed $\varepsilon > 0$ and all n.

By 1.28, any pair of points $x, y \in \mathcal{X}$ can be connected by a minimizing geodesic $[xy]$. Choose $\bar{x}_n$ on a geodesic $[x_\infty x_n]$ such that $|x_\infty - \bar{x}_n| = \varepsilon$. Note that

$$f_{x_n}(x_\infty) - f_{x_n}(\bar{x}_n) = \varepsilon,$$
$$f_{x_\infty}(x_\infty) - f_{x_n}(\bar{x}_n) = -\varepsilon$$

for all n.

Since $\mathcal{X}$ is proper, we can pass to a subsequence of x_n so that the sequence $\bar{x}_n$ converges; denote its limit by $\bar{x}_\infty$. The above identities imply that

$$f_{x_n}(\bar{x}_\infty) \not\to f_{x_\infty}(\bar{x}_\infty) \quad \text{or}$$
$$f_{x_n}(x_\infty) \not\to f_{x_\infty}(x_\infty)$$

—a contradiction.

For the second part, take $\mathcal{Y}$ to be the set of nonnegative integers with the metric ρ defined by

$$\rho(m, n) = m + n$$

for $m \neq n$.

Source I learned this example from Linus Kramer and Alexander Lytchak; it was also mentioned in the lectures of Anders Karlsson and attributed to Uri Bader [32, 2.3].

2.27. Suppose that our metric is $\sum a_S \cdot \delta_S$ with $a_S \geqslant 0$ for any $S \subset F$. Enumerate all the subsets $S_1, \ldots, S_{2^n}$; set $S_i = F$ for all $i > 2^n$. Consider the maps $x \mapsto (a_1, a_2, \ldots)$ where $a_i = 0$ if $x \in S_i$ and otherwise $a_i = 1$. Observe that it defines a distance-preserving map $F \to \ell^1$.

The if part is proved. For the only-if part, check the statement for subsets of the real line, and use it.

2.28. Show that for any proper subset S in the vertex set there are three vertices x, y, z such that $|x - y| + |y - z| = |x - z|$ and either $x, z \in S$ and $y \notin S$, or $x, z \notin S$ and $y \in S$. Then apply 2.27.

2.29. For the first part, show and use that the quotient of $\mathbb{R}\mathrm{P}^2$ by the isotropy group of one point is isometric to a line segment.

For the second part, choose three points on a closed geodesic at equal distances from each other. Show and use that there is an isometric three-point set in $\mathbb{RP}^2$ that does not lie on a closed geodesic.

Source Herbert Busemann [14, V §2].

2.30. Denote by $\dim(x_1, \dots, x_m)$ the dimension of the minimal face of the cube that contains all the points $x_1, \dots, x_m \in Q$. Show and use that

$$\dim(x_1, \dots, x_m) = \dim(x'_1, \dots, x'_m)$$

for any isometry $x \mapsto x'$ of Q.

Source Valerii Berestovskii and Yurii Nikonorov [5, prop. 6 and 7].

3.1; *Only-If Part* To check convexity, assume that B is a two-point subset. For closeness, assume that B is a countable set of A.

If Part Apply the Kirszbraun theorem together with the closest-point projection.

3.3. Choose an injective space $\mathcal{Y}$.

(a) Fix a Cauchy sequence x_n in $\mathcal{Y}$; we need to show that it has a limit $x_\infty \in \mathcal{Y}$. Consider metric on $\mathcal{X} = \mathbb{N} \cup \{\infty\}$ defined by

$$|m - n|_{\mathcal{X}} := |x_m - x_n|_{\mathcal{Y}},$$
$$|m - \infty|_{\mathcal{X}} := \lim_{n\to\infty} |x_m - x_n|_{\mathcal{Y}}.$$

Since the sequence is Cauchy, so is the sequence $\ell_n = |x_m - x_n|_{\mathcal{Y}}$ for any m. Therefore, the last limit is defined.

By construction, the map $n \mapsto x_n$ is distance-preserving on $\mathbb{N} \subset \mathcal{X}$. Since $\mathcal{Y}$ is injective, this map can be extended to ∞ as a short map; set $\infty \mapsto x_\infty$. Since $|x_n - x_\infty|_{\mathcal{Y}} \leqslant |n - \infty|_{\mathcal{X}}$ and $|n - \infty|_{\mathcal{X}} \to 0$, we get that $x_n \to x_\infty$ as $n \to \infty$.

(b) Applying the definition of injective space, we get a midpoint for any pair of points in $\mathcal{Y}$. By (a), $\mathcal{Y}$ is a complete space. It remains to apply Menger's lemma (1.27(*b*)).

(c) Let $k\colon \mathcal{Y} \hookrightarrow \ell^\infty(\mathcal{Y})$ be the Kuratowski embedding (2.4). Observe that $\ell^\infty(\mathcal{Y})$ is contractible; in particular, there is a homotopy $k_t\colon \mathcal{Y} \hookrightarrow \ell^\infty(\mathcal{Y})$ such that $k_0 = k$ and k_1 is a constant map. (In fact, one can take $k_t = (1 - t) \cdot k$.)

Since k is distance-preserving and $\mathcal{Y}$ is injective, there is a short map $f\colon \ell^\infty(\mathcal{Y}) \to \mathcal{Y}$ such that the composition $f \circ k$ is the identity map on $\mathcal{Y}$. The composition $f \circ k_t\colon \mathcal{Y} \hookrightarrow \mathcal{Y}$ provides the needed homotopy.

3.4. By 2.4, the space $\mathcal{Y}$ can be considered as a subset in $\ell^\infty(\mathcal{Y})$. Let $r\colon \ell^\infty(\mathcal{Y}) \to \mathcal{Y}$ be the short retraction provided by the definition of injective space. Observe that $m(x, y) := r(\frac{x+y}{2})$ meets the condition.

Remark The same argument can be used to construct the so-called geodesic bicombing on injective space—a useful tool introduced by Urs Lang [38, 3.6].

3.5. Suppose that a short map $f\colon A\to\mathcal{Y}$ is defined on a subset A of a metric space $\mathcal{X}$. We need to construct a short extension F of f. Without loss of generality, we may assume that $A\neq\varnothing$; otherwise, map the whole $\mathcal{X}$ to a single point. By Zorn's lemma, it is sufficient to enlarge A by a single point $x\notin A$.
(a) Suppose $\mathcal{Y}=\mathbb{R}$. Set

$$F(x)=\inf\big\{\,f(a)-|a-x|\;:\;a\in A\,\big\}.$$

Observe that F is short and $F(a)=f(a)$ for any $a\in A$.
(b) Suppose $\mathcal{Y}$ is a complete metric tree. Fix points $p\in\mathcal{X}$ and $q\in\mathcal{Y}$. Given a point $a\in A$, let $x_a\in\overline{\mathrm{B}}[f(a),|a-p|]$ be the point closest to $f(x)$. Note that $x_a\in[q\,f(a)]$ and either $x_a=q$ or x_a lies on distance $|a-p|$ from $f(a)$.

Note that the geodesics $[q\,x_a]$ are nested; that is, for any $a,b\in A$, we have either $[q\,x_a]\subset[q\,x_b]$ or $[q\,x_b]\subset[q\,x_a]$. Moreover, in the first case, we have $|x_b-f(a)|\leqslant|p-a|$, and in the second $|x_a-f(b)|\leqslant|p-b|$.

It follows that the closure of the union of all geodesics $[q\,x_a]$ for $a\in\mathcal{A}$ is a geodesic. Denote by x its endpoint; it exists since $\mathcal{Y}$ is complete. It remains to observe that $|x-f(a)|\leqslant|p-a|$ for any $a\in\mathcal{A}$; that is, one can take $f(p)=x$.
(c) Show and use that *any ℓ^∞-product of injective spaces is injective*; in particular, if $\mathcal{Y}$ and $\mathcal{Z}$ are injective, then so is the product $\mathcal{Y}\times\mathcal{Z}$ equipped with the metric

$$|(y,z)-(y',z')|_{\mathcal{Y}\times\mathcal{Z}}=\max\{\,|y-y'|_{\mathcal{Y}},|z-z'|_{\mathcal{Z}}\,\}.$$

3.6; (a) Let $\mathcal{B}=\overline{\mathrm{B}}[o,R]_{\mathcal{Y}}$. Choose a metric space $\mathcal{X}$ with a subset A. Given a short map $f\colon A\to\mathcal{B}$, we need to find its short extension $\mathcal{X}\to\mathcal{B}$.

Since $\operatorname{diam}\mathcal{B}\leqslant 2\cdot R$, we may assume that $\operatorname{diam}\mathcal{X}\leqslant 2\cdot R$; if not pass to the metric defined by $|x-y|:=\max\{\,|x-y|_{\mathcal{X}},2\cdot R\,\}$.

Let us add point w to $\mathcal{X}$ such that $|w-x|=R$ for any $x\in\mathcal{X}$; denote the obtained space $\mathcal{X}'$. Let $f'\colon A\cup\{w\}\to\mathcal{B}$ be an extension of f by $w\mapsto o$; note that f' is short.

Since $\mathcal{Y}$ is injective, there is a short extension $F\colon\mathcal{X}'\to\mathcal{Y}$ of f'. Show and use that $F(\mathcal{X}')\subset\mathcal{B}$.
(b) Try to modify the argument in (a). Namely, let $\mathcal{B}=\bigcap_\alpha\overline{\mathrm{B}}[o_\alpha,R_\alpha]_{\mathcal{Y}}$. Note that one may assume that $\operatorname{diam}\mathcal{X}\leqslant 2\cdot\inf_\alpha\{\,R_\alpha\,\}$. Consider the space $\mathcal{X}'=\mathcal{X}\cup\{w_\alpha\}$ such that $|w_\alpha-x|=R_\alpha$ for any $x\in\mathcal{X}$ and $|w_\alpha-w_\beta|=R_\alpha+R_\beta$ if $\alpha\neq\beta$. Further, consider an extension of f by $w_\alpha\mapsto o_\alpha$.
3.7. Let $\operatorname{diam}\mathcal{Y}=2\cdot R$. We can assume that $R>0$; otherwise, there is nothing to prove. Denote by $\mathcal{Z}$ a minimal (with respect to inclusion) intersection of closed R-balls in $\mathcal{Y}$ such that $s(\mathcal{Z})\subset\mathcal{Z}$.

Consider the intersection

$$\mathcal{Y}'=\mathcal{Z}\cap\left(\bigcap_{p\in\mathcal{Z}}\overline{\mathrm{B}}[p,R]_{\mathcal{Y}}\right).$$

By 3.6(b), $\mathcal{Y}'$ is injective. Use that $\mathcal{Z}$ is minimal to show that $s(\mathcal{Y}') \subset \mathcal{Y}'$. Show that $\operatorname{diam}\mathcal{Y}' \leqslant \frac{1}{2} \cdot \operatorname{diam}\mathcal{Y}$.

Consider a sequence of nested injective spaces $\mathcal{Y} = \mathcal{Y}_0 \supset \mathcal{Y}_1 \supset \dots$ such that $\mathcal{Y}_{n+1} = \mathcal{Y}'_n$. Choose a point $y_n \in \mathcal{Y}_n$ for each n. Show that the sequence y_n is Cauchy. By 3.3(a), y_n converges, say to y_∞. Observe that y_∞ is a fixed point of s.

3.12; *Only-If Part* Suppose r is extremal. By 3.11(b), r is 1-Lipschitz. Since $\mathbb{S}^1$ is compact, 3.11(b) implies that for any $p \in \mathbb{S}^1$, there is $q \in \mathbb{S}^1$ such that

$$r(p) + r(q) = |p - q|_{\mathbb{S}^1}.$$

Therefore,

$$\begin{aligned}\pi &= |p - (-p)|_{\mathbb{S}^1} \leqslant \\ &\leqslant r(p) + r(-p) = \\ &= r(p) + r(q) + r(-p) - r(q) \leqslant \\ &\leqslant |p - q|_{\mathbb{S}^1} + |q - (-p)|_{\mathbb{S}^1} = \\ &= \pi.\end{aligned}$$

So, we have equality in both places, and the only-If part follows.

If Part Assume r is a 1-Lipschitz function such that $r(p) + r(-p) = \pi$. Then

$$\begin{aligned}|p - q|_{\mathbb{S}^1} &= |p - (-p)|_{\mathbb{S}^1} - |q - (-p)|_{\mathbb{S}^1} \geqslant \\ &\geqslant \pi - (r(-p) - r(q)) = \\ &= r(p) + r(q).\end{aligned}$$

Therefore, r is admissible.

Finally, if r is not extremal, then there is an admissible function $s \leqslant r$ such that $s(p) < r(p)$ for some p. The latter contradicts the equality $r(p) + r(-p) = \pi$.

Source Roger Züst [68, Proposition 2.7].

3.13. Show and use that $s^*(x) \geqslant |x - y| - s(y)$ for any $x, y \in \mathcal{X}$.

Remarks It is easy to check that $q\colon s \mapsto \frac{1}{2} \cdot (s + s^*)$ is a short map on the space of admissible functions (with sup-norm). Moreover, iterating q and passing to the limit, we get a short retraction from the space of admissible functions to the space of extremal functions on $\mathcal{X}$ [see 3.1 in 38]. The existence of such a map also follows from 3.27.

3.15. Apply 3.14(c).

Comment Conditions under which gluings of injective spaces are injective were studied by Benjamin Miesch and Maël Pavón [44, 45].

3.16. Let $B = \overline{\mathrm{B}}[0, 1]$ and $P \supset B$ be a parallelepiped of minimal volume. Choose coordinates so that P is described by inequalities $|x_i| \leqslant 1$ for all i; let $e_1, \dots, e_m$ be the standard basis of these coordinates.

Let $B_i = \overline{\mathrm{B}}[(1 + R) \cdot e_i, R]$ for some $R > 0$. Show that $e_i \in B$ for any i; in particular, $B \cap B_i \neq \varnothing$. Show P can be chosen so that $B_i \cap B_j \neq \varnothing$ for all i and j and all large $R > 0$. Apply hyperconvexity to show that $e_1 + \dots + e_m \in B$. The same way, show that $\pm e_1 \pm \dots \pm e_m \in B$ for all choices of signs. Conclude that $B = P$.

3.17. Observe that closed balls are compact and apply the finite intersection property.

3.18. Denote by $\mathcal{U}_d$ the d-Urysohn space, so $\mathcal{U}_\infty$ is the Urysohn space.

The extension property implies finite hyperconvexity. It remains to show that $\mathcal{U}_d$ is not countably hyperconvex.

Suppose that $d < \infty$. Show that for any point $x \in \mathcal{U}_d$ there is a point $y \in \mathcal{U}_d$ such that $|x - y|_{\mathcal{U}_d} = d = \operatorname{diam}\mathcal{U}_d$. It follows that there is no point $z \in \mathcal{U}_d$ such that $|z - x|_{\mathcal{U}_d} \leqslant \frac{d}{2}$ for any $x \in \mathcal{U}_d$. Whence $\mathcal{U}_d$ is not countably hyperconvex.

Use 2.19(*b*) to reduce the case $d = \infty$ to the case $d < \infty$.

3.19. Choose $\varepsilon_0 > 0$. Let p_0 be a point provided by the definition of almost hyperconvexity; that is, $|x_\alpha - p_0| \leqslant r_\alpha + \varepsilon_0$ for a given $\varepsilon_0 > 0$. We may assume that $\delta_0 = \sup\{\, |x_\alpha - p_0| - r_\alpha \,\} > 0$; otherwise, the problem is solved. Clearly, $\delta_0 \leqslant \varepsilon_0$.

Applying hyperconvexity for $\varepsilon_1 < \frac{1}{10} \cdot \delta_0$, we get a point p_1 such that $|x_\alpha - p_1| \leqslant r_\alpha + \varepsilon_1$ and $|p_0 - p_1| \leqslant \delta_0 + \varepsilon_1$. Again, we may assume that $\delta_1 = \sup\{\, |x_\alpha - p_1| - r_\alpha, |p_0 - p_1| \,\} > 0$, and we have $\delta_1 \leqslant \varepsilon_1$.

Continuing this way, we get a sequence $p_0, p_1, \dots$ that either terminates and in this case the problem is solved, or it is an infinite Cauchy sequence. In the latter case, its limit p_∞ satisfies $|x_\alpha - p_\infty| \leqslant r_\alpha$ for any α.

Comment This solution reminds the proof of 2.9; a more exact statement was proved by Benjamin Miesch and Maël Pavón [46, 2.2]; namely, they show that almost n-hyperconvexity implies $(n - 1)$-hyperconvexity.

3.20. Show and use that the functions in $\operatorname{Ext}\mathcal{X}$ are 1-Lipschitz and uniformly bounded.

3.21; (a) Use 3.11(*d*) to show that if f is extremal if and only if $f(v) = x$ and $f(w) = 1 - x$ for some $x \in [0, 1]$. Conclude that $\operatorname{Ext}\mathcal{X}$ is isometric to the unit interval $[0, 1]$.

(b) Let f be an extremal function. By 3.11(*d*), at least two of the numbers $f(a) + f(b)$, $f(b)+f(c)$, and $f(c)+f(a)$ are 1. It follows that for some $x \in [0, \frac{1}{2}]$, we have

$$f(a) = 1 \pm x, \qquad f(b) = 1 \pm x, \qquad f(c) = 1 \pm x,$$

where we have one "minus" and two "pluses" in these three formulas.

Suppose that

$$g(a) = 1 \pm y, \qquad g(b) = 1 \pm y, \qquad g(c) = 1 \pm y$$

is another extremal function. Then $|f - g| = |x - y|$ if g has "minus" at the same place as f and $|f - g| = |x + y|$ otherwise.

It follows that $\operatorname{Ext} \mathcal{X}$ is isometric to a *tripod*—three segments of length $\frac{1}{2}$ glued at one end.

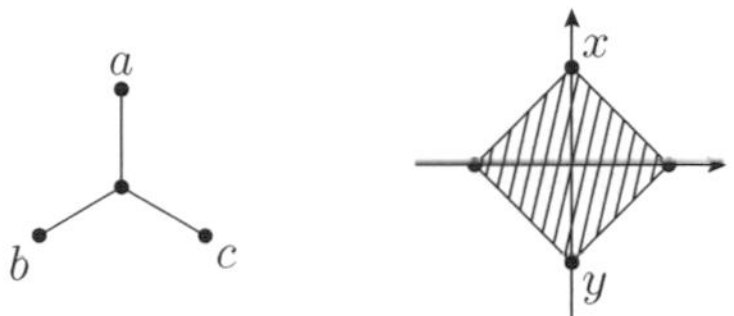

(c) Assume f is an extremal function. Use 3.11((*d*)) to show that

$$\begin{aligned} 2 &= f(x) + f(y) = \\ &= f(p) + f(q); \end{aligned}$$

in particular, two values $a = f(x) - 1$ and $b = f(p) - 1$ completely describe the function f. Since f is extremal, we also have that

$$(1 \pm a) + (1 \pm b) \geqslant 1$$

for all 4 choices of signs; equivalently,

$$|a| + |b| \leqslant 1.$$

It follows that $\operatorname{Ext} \mathcal{X}$ is isometric to the rhombus $|a| + |b| \leqslant 1$ in the (a, b)-plane with the metric induced by the ℓ^∞-norm.

Remarks If $\mathcal{X}$ has n-points, then (evidently) $\operatorname{Ext} \mathcal{X}$ is a polyhedral complex in $(\mathbb{R}^n, \ell^\infty) = \ell^\infty(\mathcal{X})$; each face of the complex is defined by equalities and inequalities of the following type: $x_i + x_j \geqslant \text{const}$ and $x_i + x_j = \text{const}$. It is easy to see (and follows from 3.16) that each face is isometric to a convex polyhedron in $(\mathbb{R}^k, \ell^\infty)$ for some $k \leqslant n$; in fact $k \leqslant n/2$. The structure of the complex can be encoded by certain graphs with the vertex set $\mathcal{X}$ [see Section 4 in 38].

3.22. Recall that $x \mapsto \operatorname{dist}_x$ gives an isometric embedding $\mathcal{X} \hookrightarrow \ell^\infty(\mathcal{X})$; so we can identify $\mathcal{X}$ with a subset of $\ell^\infty(\mathcal{X})$. Further, $\operatorname{Ext} \mathcal{X}$ is a subset of $\ell^\infty(\mathcal{X})$. It is sufficient to show that $\operatorname{Ext} \mathcal{X} = G$.

Use 3.11(*d*) to show that $\operatorname{Ext} \mathcal{X} \subset G$.

Given $g \in G$, show that $g(x) = |g - x|_{\ell^\infty(\mathcal{X})}$. Conclude that g is admissible and apply 3.11(*d*).

Source Suggested by Rostislav Matveyev.

3.25. Recall that

$$|f - g|_{\operatorname{Ext}\mathcal{X}} = \sup\{\,|f(x) - g(x)| \,:\, x \in \mathcal{X}\,\}$$

and

$$|f - p|_{\operatorname{Ext}\mathcal{X}} = f(p)$$

for any $f, g \in \operatorname{Ext}\mathcal{X}$ and $p \in \mathcal{X}$.

Since $\mathcal{X}$ is compact, we can find a point $p \in \mathcal{X}$ such that

$$\begin{aligned}|f - g|_{\operatorname{Ext}\mathcal{X}} &= |f(p) - g(p)| = \\ &= \big||f - p|_{\operatorname{Ext}\mathcal{X}} - |g - p|_{\operatorname{Ext}\mathcal{X}}\big|.\end{aligned}$$

Without loss of generality, we may assume that

$$|f - p|_{\operatorname{Ext}\mathcal{X}} = |g - p|_{\operatorname{Ext}\mathcal{X}} + |f - g|_{\operatorname{Ext}\mathcal{X}}.$$

Applying 3.11(*d*), we can find a point $q \in \mathcal{X}$ such that

$$|q - p|_{\operatorname{Ext}\mathcal{X}} = |f - p|_{\operatorname{Ext}\mathcal{X}} + |f - q|_{\operatorname{Ext}\mathcal{X}},$$

whence the result.

Since $\operatorname{Ext}\mathcal{X}$ is injective (3.23), by 3.3(*b*), it has to be geodesic. It remains to note that the concatenation of geodesics $[pq]$, $[gf]$, and $[fq]$ is the required geodesic $[pq]$.

3.26. The only-if part follows since $\mathcal{X}$ is isometric to a subset of $\operatorname{Ext}\mathcal{X}$.

The if part means that

$$\text{①}\quad \begin{aligned}|f - g| + |v - w| \leqslant \max\{\,&|f - v| + |g - w|,\\ &|f - w| + |g - v|\,\} + 2\cdot\delta\end{aligned}$$

for any $f, g, v, w \in \operatorname{Ext}\mathcal{X}$.

Suppose $\mathcal{X}$ is compact. Applying 3.25, we can choose $p, q, x, y \in \mathcal{X}$ such that

$$\text{②}\quad \begin{aligned}|p - f| + |f - g| + |g - q| &= |p - q|\\ |x - v| + |v - w| + |w - y| &= |x - y|\end{aligned}.$$

Since $\mathcal{X}$ is δ-hyperbolic, we have

$$\begin{aligned}|p - q| + |x - y| \leqslant \max\{\,&|p - x| + |q - y|,\\ &|p - y| + |q - x|\,\} + 2\cdot\delta.\end{aligned}$$

Show that this inequality, together with the triangle inequality, and ② imply ①.

For the noncompact case, prove an approximate version of ② and apply it the same way.

3.28. Show that there is a unique isometry of Ext $\mathcal{X}$ that is identity on $\mathcal{X}$. Use it together with 3.27.

3.29. Show that there is a pair of short maps Ext $\mathcal{X} \to$ Ext $\mathcal{U} \to$ Ext $\mathcal{X}$ such that their composition is the identity on $\mathcal{X}$. You may need to apply the Katětov extension (2.5(a)). Make a conclusion.

3.30. Apply 3.11(d) to show that for any $u \in \mathbb{S}^2_+$ the restriction $f_u := \operatorname{dist}_u|_{\mathbb{S}^1}$ is an extremal function on $\mathbb{S}^1$. Moreover, the function f_u uniquely determines u. Make a conclusion.

3.32. Observe that coordinate functions are monotonic on any geodesic in ℓ^1. Use it to show that ℓ^1 is a median space; that is, for any three points x, y, z, there is a *unique* point m (it is called the median of x, y, and z) that lies on some geodesics $[xy]$, $[xz]$, and $[yz]$. Apply it to show that ℓ^1 is 3-hyperconvex.

The 4-hyperconvexity fails for the unit balls centered at four even vertices of the cube $([0, 1]^3, \ell^1)$.

3.33. Choose three points $x, y, z \in \mathcal{X}$ and set $A = \{x, z\}$. Let $f : A \to A$ be the identity map. Then $F(y) = x$ or $F(y) = z$. In both cases, the strong triangle inequality follows.

3.34; *main part* Choose a maximal (with respect to inclusion) subset $A \supset \mathcal{K}$ that admits a short retraction $f : A \to \mathcal{K}$; it exists by Zorn's lemma. If A is the whole space, then the problem is solved. Otherwise, choose $p \notin A$.

Choose a sequence of points $a_n \in A$ such that $|a_n - p|$ converge to the exact lower bound on the distances from points in A to p. Since $\mathcal{K}$ is compact, we can pass to a subsequence of a_n such that $f(a_n)$ converges. Let

$$f(p) = \lim f(a_n).$$

It remains to check that

$$③ \qquad |f(a) - f(p)| \leqslant |a - p|$$

for any $a \in A$. Choose $\varepsilon > 0$; note that

$$|a_n - p| < |a - p| + \varepsilon, \qquad |f(a_n) - f(p)| < \varepsilon$$

for all large n. Therefore,

$$\begin{aligned}|f(a) - f(p)| &\leqslant \\ &\leqslant \max\{\,|f(a) - f(a_n)|, |f(a_n) - f(p)|\,\} \leqslant \\ &\leqslant |f(a) - f(a_n)| + \varepsilon \leqslant \\ &\leqslant |a - a_n| + \varepsilon \leqslant\end{aligned}$$

$$\leqslant \max\{\,|a-p|, |a_n - p|\,\} + \varepsilon <$$
$$< |a-p| + 2\cdot\varepsilon.$$

Since $\varepsilon > 0$ is arbitrary, we get ③.

Example Consider set of $\{\infty, 1, 2, \dots\}$ with metric defined by

$$|m-n| := 1 + \frac{1}{\min\{m,n\}}$$

for $m \neq n$. Observe that the space is complete, the subset $\{1, 2, \dots\}$ is closed, but it is not a short retract of the ambient space.

3.35. Consider the space $\mathcal{K}^{\mathcal{X}}$ of all maps $\mathcal{X} \to \mathcal{K}$ equipped with the product topology.

Denote by $\mathfrak{S}_F$ the set of maps $h \in \mathcal{K}^{\mathcal{X}}$ such that the restriction $h|_F$ is short and agrees with f in $F \cap A$. Note that the sets $\mathfrak{S}_F \subset \mathcal{K}^{\mathcal{X}}$ are closed and any finite intersection of these sets is nonempty.

According to Tikhonov's theorem, $\mathcal{K}^{\mathcal{X}}$ is compact. By the finite intersection property, the intersection $\bigcap_F \mathfrak{S}_F$ for all finite sets $F \subset X$ is nonempty. Hence, the statement follows.

Source Stephan Stadler and the author [53, 7.1].

4.3. Suppose that $|A - B|_{\mathrm{Haus}\mathcal{X}} < r$. Choose a pair of points $a, a' \in A$ on maximal distance from each other. Observe that there are points $b, b' \in B$ such that $|a-b|_{\mathcal{X}}, |a'-b'|_{\mathcal{X}} < r$. Whence

$$|a-a'|_{\mathcal{X}} - |b-b'|_{\mathcal{X}} \leqslant 2\cdot r,$$

and therefore,

$$\operatorname{diam} A - \operatorname{diam} B \leqslant 2\cdot |A-B|_{\mathrm{Haus}\mathcal{X}}.$$

Swap A and B and repeat the argument.

4.4; ((a)) Given a set $A \subset \mathbb{R}^2$, denote by A^r its closed r-neighborhood. Show and use that

$$(\operatorname{Conv} A)^r = \operatorname{Conv}(A^r).$$

((b)) The answer is "no" in both parts.

For the first part, let A be a unit disk and B a finite ε-net in A. Evidently, $|A-B|_{\mathrm{Haus}\mathbb{R}^2} < \varepsilon$, but $|\partial A - \partial B|_{\mathrm{Haus}\mathbb{R}^2} \approx 1$.

For the second part, take A to be a unit disk and $B = \partial A$ to be its boundary circle. Note that $\partial A = \partial B$; in particular, $|\partial A - \partial B|_{\mathrm{Haus}\mathbb{R}^2} = 0$, while $|A-B|_{\mathrm{Haus}\mathbb{R}^2} = 1$.

Remark A more interesting example for (b) is provided by the so-called *lakes of Wada*—an example of three (and more) disjoint open topological disks in the plane that have identical boundaries.

4.5. Checking two functions dist_A and dist_B leads to

$$|A - B| \leqslant \sup_f \{ \max_{a\in A}\{f(a)\} - \max_{b\in B}\{f(b)\} \}.$$

Use 4.2 to prove the opposite inequality.
4.6. By 4.4(a), it is sufficient to show that

$$|A - B|_{\mathrm{Haus}\mathbb{R}^n} = \sup_{|u|=1} \{|h_A(u) - h_B(u)|\}$$

for any nonempty compact convex sets $A, B \subset \mathbb{R}^n$.

Prove the 1-dimensional case of this equality. Further, denote by A_ℓ the orthogonal projection of A to a line ℓ. Show and use that

$$|A - B|_{\mathrm{Haus}\mathbb{R}^n} = \sup_\ell\{|A_\ell - B_\ell|_{\mathrm{Haus}\ell}\},$$

where the least upper bound is taken for all lines ℓ.
4.7; (a) Given $t \in (0, 1]$, consider the real interval $\tilde{C}_t = [\frac{1}{t} + t, \frac{1}{t} + 1]$. Denote by C_t the image of $\tilde{C}_t$ under the covering map $\pi\colon \mathbb{R} \to \mathbb{S}^1 = \mathbb{R}/\mathbb{Z}$.

Set $C_0 = \mathbb{S}^1$. Note that the Hausdorff distance from C_0 to C_t is $\frac{t}{2}$. Therefore, $\{C_t\}_{t\in[0,1]}$ is a family of compact subsets in $\mathbb{S}^1$ that is continuous in the sense of Hausdorff.

Assume there is a continuous section $c(t) \in C_t$, for $t \in [0, 1]$. Since π is a covering map, we can lift the path c to a path $\tilde{c}\colon [0, 1] \to \mathbb{R}$ such that $\tilde{c}(t) \in \tilde{C}_t$ for all t. In particular, $\tilde{c}(t) \to \infty$ as $t \to 0$, a contradiction.
(b) Consider path $c(t) := \min C_t$.

Source Suggested by Stephan Stadler.

4.9. Show that $f\colon (p,r) \mapsto \overline{\mathrm{B}}[p,r]$ defines a continuous map $\mathcal{X} \times [0,\infty) \to \mathrm{Haus}\mathcal{X}$. Observe that $\overline{\mathrm{B}}[p,r] = \mathcal{X}$ for all $r \geqslant \mathrm{diam}\mathcal{X}$. Composing f with the retraction $r\colon \mathrm{Haus}\mathcal{X} \to \mathcal{X}$, we get a homotopy from the identity map to a constant map on $\mathcal{X}$, hence the statement.
By the way, is there a good description of such spaces? Note that if $\mathcal{X}$ is injective or discrete, then a short retraction $\mathrm{Haus}\mathcal{X} \to \mathcal{X}$ exists. On the other hand, the Euclidean plane does not have such a retraction [49, 58]. See also [52].
4.12. Show that for any $\varepsilon > 0$ there is a positive integer N such that $\bigcup_{n\leqslant N} K_n$ is an ε-net in the union $\bigcup_n K_n$. Observe that $\bigcup_{n\leqslant N} K_n$ is compact. Apply 1.13 and 1.11(c).
4.13; *if part* Choose two compact sets $A, B \subset \mathcal{X}$; suppose that $|A - B|_{\mathrm{Haus}\mathcal{X}} < r$.

Choose finite ε-nets $\{a_1, \ldots, a_m\} \subset A$ and $\{b_1, \ldots b_n\} \subset B$. For each pair a_i, b_j, construct a constant-speed path $\gamma_{i,j}$ from a_i to b_j such that

$$\operatorname{length}\gamma_{i,j} < |a_i - b_j| + \varepsilon.$$

Set

$$C(t) = \left\{ \gamma_{i,j}(t) \ : \ |a_i - b_j|_{\mathcal{X}} < r + \varepsilon \right\}.$$

Observe that $C(t)$ is finite; in particular, it is compact.

Show and use that

$$|A - C(t)|_{\mathcal{X}} < t \cdot r + 10 \cdot \varepsilon,$$

$$|C(t) - B|_{\mathcal{X}} < (1 - t) \cdot r + 10 \cdot \varepsilon.$$

Apply 4.12 and 1.27.

Only-If Part Choose points $p, q \in \mathcal{X}$. Show that the existence of ε-midpoints between $\{p\}$ and $\{q\}$ in $\operatorname{Haus}\mathcal{X}$ implies the existence of ε-midpoints between p and q in $\mathcal{X}$. Apply 1.27.

4.14; (a) Suppose that a sequence of compact subsets $K_n \subset \mathbb{R}^2$ converges to K_∞ in the sense of Hausdorff. Assume K_∞ is not connected; show that so is K_n for large n.
(b) Choose a finite subset F that is ε-close to K. Show that one can obtain a tree T by connecting some vertices of F by line segments of length smaller than $2 \cdot \varepsilon$. Let γ be a curve that bounds a neighborhood of T. Show that if the neighborhood is sufficiently small, then γ is $2 \cdot \varepsilon$-close to K.

Remarks You might be surprised to learn that most connected compact sets in the pane are homeomorphic to each other—they are homeomorphic to the so-called pseudo-arc [7]; here the word *most* understood in the sense of F. In particular, most of the compact connected plane sets are not path-connected.

4.15. Let A be a compact convex set in the plane. Denote by A^r the closed r-neighborhood of A. Recall that by Steiner's formula, we have

$$\operatorname{area}A^r = \operatorname{area}A + r \cdot \operatorname{perim}A + \pi \cdot r^2.$$

Taking the derivative and applying the coarea formula, we get

$$\operatorname{perim}A^r = \operatorname{perim}A + 2 \cdot \pi \cdot r.$$

Observe that if A lies in a compact set B bounded by a closed curve, then

$$\operatorname{perim}A \leqslant \operatorname{perim}B.$$

Indeed, the closest-point projection $\mathbb{R}^2 \to A$ is short, and it maps ∂B onto ∂A.

It remains to use the following observation: if $A_n \to A_\infty$, then for any $r > 0$, we have that the inclusions

$$A_\infty^r \supset A_n \quad \text{and} \quad A_\infty \subset A_n^r$$

hold for all large n.

4.17. Note that almost all points on ∂D have a defined tangent line. In particular, for almost all pairs of points $a, b \in \partial D$, the two angles α and β between the chord $[ab]$ and ∂D are defined.

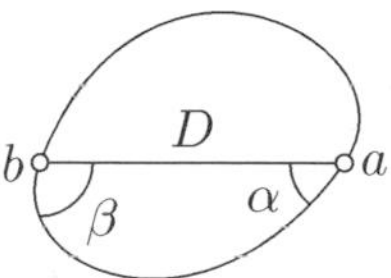

The convexity of D' implies that $\alpha = \beta$; here we measure the angles α and β on one side from $[ab]$. Show that if the identity $\alpha = \beta$ holds for almost all chords, then D is a round disk.

4.19. Observe that all functions $\operatorname{dist}_{A_n}$ are Lipschitz. Suppose that for some (and therefore any) point x, the sequence $\operatorname{dist}_{A_n}(x)$ is not bounded. Then we can pass to a subsequence of A_n so that $\operatorname{dist}_{A_n}(x) \to \infty$ for any x; in this case, A_n converges to the empty set.

Assume the sequence $\operatorname{dist}_{A_n}(x)$ is bounded for some (and therefore any) point x. Then, passing to a subsequence of A_n, we may assume that the sequence $\operatorname{dist}_{A_n}$ converges to some function f.

Set $A_\infty = f^{-1}\{0\}$. It remains to show that $f = \operatorname{dist}_{A_\infty}$.

5.3; (a) Apply the definition (5.1) for space $\mathcal{W}$ obtained from $\mathcal{X}$ by adding one point that lies at distance $\frac{1}{2} \cdot \operatorname{diam}\mathcal{X}$ from each point of $\mathcal{X}$.

(b) Given a point $x \in \mathcal{X}$, denote by $a \cdot x$ and $b \cdot x$ the corresponding points in $a \cdot \mathcal{X}$ and $b \cdot \mathcal{X}$, respectively. Show that there is a metric on $\mathcal{W} = a \cdot \mathcal{X} \sqcup b \cdot \mathcal{X}$ such that

$$|a \cdot x - b \cdot x|_{\mathcal{W}} = \tfrac{|b-a|}{2} \cdot \operatorname{diam}\mathcal{X}$$

for any x and the inclusions $a \cdot \mathcal{X} \hookrightarrow \mathcal{W}$, $b \cdot \mathcal{X} \hookrightarrow \mathcal{W}$ are distance-preserving. Conclude that $|\mathcal{X} - \mathcal{O}|_{\mathrm{GH}} \leqslant \frac{1}{2} \cdot \operatorname{diam}\mathcal{X}$. The opposite inequality follows from (a).

(c) Use (a) and (b) to show that the isometry class of $\mathcal{O}$ is completely determined by the following property:

$$|\mathcal{X} - \mathcal{Y}|_{\mathrm{GH}} \leqslant \max\{\, |\mathcal{O} - \mathcal{X}|_{\mathrm{GH}}, |\mathcal{O} - \mathcal{Y}|_{\mathrm{GH}} \,\},$$

for any $\mathcal{X}$ and $\mathcal{Y}$.

Remark In fact, *the isometry group of space* GH *is trivial*. The latter was proved by George Lowther [31, 41].

5.4. Check a one-point set and the vertices of an equilateral triangle. You may use 5.3(a).

5.5. Suppose that we can identify $\mathcal{A}_r$ and $\mathcal{B}_r$ with subspaces of a space $\mathcal{W}$ such that

$$|\mathcal{A}_r - \mathcal{B}_r|_{\mathrm{Haus}\mathcal{W}} < \tfrac{1}{10}$$

for large r; see the definition of Gromov–Hausdorff metric (5.1).

Set $n = \lceil r \rceil$. Note that there are $2 \cdot n$ integer points in $\mathcal{A}_r$: $a_1 = (0,0)$, $a_2 = (1,0)$, $\dots$, $a_{2\cdot n} = (n,1)$. Choose a point $b_i \in \mathcal{B}_r$ that lies at the minimal distance from a_i. Note that $|b_i - b_j| > \tfrac{4}{5}$ if $i \neq j$. It follows that $r > \tfrac{4}{5} \cdot (2 \cdot n - 1)$. The latter contradicts $n = \lceil r \rceil$ for large r.

Remark Try to show that $|\mathcal{A}_r - \mathcal{B}_r|_{\mathrm{GH}} = \tfrac{1}{2}$ for all large r.

5.6. Suppose

$$|\mathcal{X} - \mathcal{Y}|_{\mathrm{Haus}\mathcal{U}} < \varepsilon.$$

Denote by $\hat{\mathcal{U}}$ the injective envelope of $\mathcal{U}$. According to 3.29, the inclusions $\mathcal{X} \hookrightarrow \mathcal{U}$ and $\mathcal{Y} \hookrightarrow \mathcal{U}$ can be extended to distance-preserving inclusions $\hat{\mathcal{X}} \hookrightarrow \hat{\mathcal{U}}$ and $\hat{\mathcal{Y}} \hookrightarrow \hat{\mathcal{U}}$. Therefore, we can and will consider $\hat{\mathcal{X}}$ and $\hat{\mathcal{Y}}$ as subspaces of $\hat{\mathcal{U}}$. It is sufficient to show that

① $$|\hat{\mathcal{X}} - \hat{\mathcal{Y}}|_{\mathrm{Haus}\hat{\mathcal{U}}} < 2 \cdot \varepsilon.$$

Given $f \in \hat{\mathcal{U}}$, let us find $g \in \hat{\mathcal{X}}$ such that

② $$|f(u) - g(u)| < 2 \cdot \varepsilon$$

for any $u \in \mathcal{U}$. Note that the restriction $f|_{\mathcal{X}}$ is admissible on $\mathcal{X}$. By 3.9, there is $g \in \hat{\mathcal{X}}$ such that

③ $$g(x) \leqslant f(x)$$

for any $x \in \mathcal{X}$.

Recall that any extremal function is 1-Lipschitz; in particular, f and g are 1-Lipschitz on $\mathcal{U}$. Therefore, (3) and $|\mathcal{X} - \mathcal{Y}|_{\mathcal{U}} < \varepsilon$ imply that

$$g(u) < f(u) + 2 \cdot \varepsilon$$

for any $u \in \mathcal{U}$. By 3.10, we also have

$$g(u) > f(u) - 2 \cdot \varepsilon$$

for any $u \in \mathcal{U}$. Whence ② follows.

It follows that $\hat{\mathcal{Y}}$ lies in a $2 \cdot \varepsilon$-neighborhood of $\hat{\mathcal{X}}$ in $\hat{\mathcal{U}}$. The same way we show that $\hat{\mathcal{X}}$ lies in a $2 \cdot \varepsilon$-neighborhood of $\hat{\mathcal{Y}}$ in $\hat{\mathcal{U}}$. Hence, ① follows.

Remark This problem was discussed by Urs Lang, Maël Pavón, and Roger Züst [39, 3.1]. They also show that the constant 2 is optimal. To see this, look at the injective envelopes of two four-point metric spaces shown on the diagram and observe that the Gromov–Hausdorff distance between the 4-point metric spaces is 1, while the distance between their injective envelopes approaches 2 as $s \to \infty$.

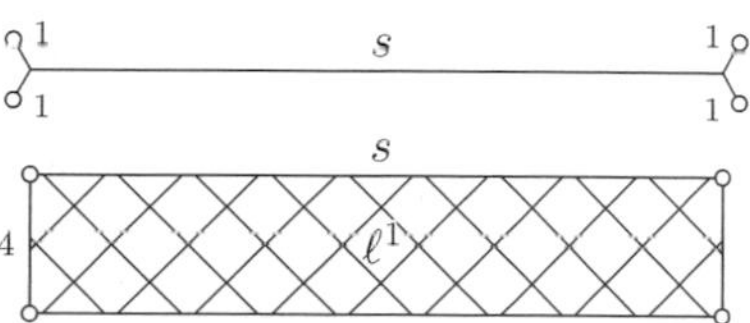

5.8; *Only-If Part* Let us identify $\mathcal{X}$ and $\mathcal{Y}$ with subspaces of a metric space $\mathcal{W}$ such that

$$|\mathcal{X} - \mathcal{Y}|_{\text{Haus}\mathcal{W}} < \varepsilon.$$

Set $x \approx y$ if and only if $|x - y|_{\mathcal{W}} < \varepsilon$. It remains to check that $\approx$ is an ε-approximation.

If Part Show that we can assume that

$$R = \{ (x, y) \in \mathcal{X} \times \mathcal{Y} : x \approx y \}$$

is a compact subset of $\mathcal{X} \times \mathcal{Y}$. Conclude that

$$\big||x - x'|_{\mathcal{X}} - |y - y'|_{\mathcal{Y}}\big| < 2 \cdot \varepsilon'$$

for some $\varepsilon' < \varepsilon$.

Show that there is a metric on $\mathcal{W} = \mathcal{X} \sqcup \mathcal{Y}$ such that the inclusions $\mathcal{X} \hookrightarrow \mathcal{W}$ and $\mathcal{Y} \hookrightarrow \mathcal{W}$ are distance-preserving and $|x - y|_{\mathcal{W}} = \varepsilon'$ if $x \approx y$. Conclude that

$$|\mathcal{X} - \mathcal{Y}|_{\text{Haus}\mathcal{W}} \leqslant \varepsilon' < \varepsilon.$$

5.10; (a) Let $\approx$ be an ε-approximation provided by 5.8. For any $x \in \mathcal{X}$, choose a point $f(x) \in \mathcal{Y}$ such that $x \approx f(x)$. Show that $x \mapsto f(x)$ is an $2 \cdot \varepsilon$-isometry.
(b) Let $x \in \mathcal{X}$ and $y \in \mathcal{Y}$. Set $x \approx y$ if $|y - f(x)|_{\mathcal{Y}} < \varepsilon$. Show that $\approx$ is an ε-approximation. Apply 5.8.
5.13. Consider the product space $[0, \varepsilon] \times \mathbb{Z}_n$ with the natural ℓ^∞-product metric. Make three variations of it by changing the sizes of some segments.

5.15; (a) Suppose $\mathcal{X}_n$ are simply-connected length metric space, $\mathcal{X}_n \xrightarrow{\text{GH}} \mathcal{X}$, and there is a nontrivial covering map $f : \tilde{\mathcal{X}} \to \mathcal{X}$. We will arrive at a contradiction by showing that there is a nontrivial covering map $f_n : \tilde{\mathcal{X}}_n \to \mathcal{X}_n$ for large n.

Choose a base point $p \in \mathcal{X}$ and its inverse image $\tilde{p} \in \tilde{\mathcal{X}}$. Consider two paths $\alpha, \alpha' : [0, 1] \to \mathcal{X}$ that start at p; denote by $\tilde{\alpha}, \tilde{\alpha}' : [0, 1] \to \tilde{\mathcal{X}}$ their liftings. Show that there is $\varepsilon > 0$ such that if $|\alpha(t) - \alpha'(t)|_{\mathcal{X}} < \varepsilon$ for any t, then $|\tilde{\alpha}(1) - \tilde{\alpha}'(1)|_{\tilde{\mathcal{X}}} < \varepsilon$.

Now suppose n is large. Choose an $\frac{\varepsilon}{10}$-approximation $\approx$ for $\mathcal{X}_n$ and $\mathcal{X}$. Choose $q \in \mathcal{X}_n$ such that $q \approx p$. Show that for any path $\beta : [0, 1] \to \mathcal{X}_n$ that starts at q, there is a path $\alpha : [0, 1] \to \mathcal{X}$ that starts at p such that $\alpha(t) \approx \beta(t)$ for any t. Observe that if α and α' are two choices of such paths, then $|\alpha(t) - \alpha'(t)|_{\mathcal{X}} < \varepsilon$.

Mimicking the standard construction of a covering map, we get the needed $f_n : \tilde{\mathcal{X}}_n \to \mathcal{X}_n$.

(b) Let $\mathcal{V}$ be a cone over Hawaiian earrings. Consider the *doubled cone* $\mathcal{W}$—two copies of $\mathcal{V}$ with glued base points (see the diagram).

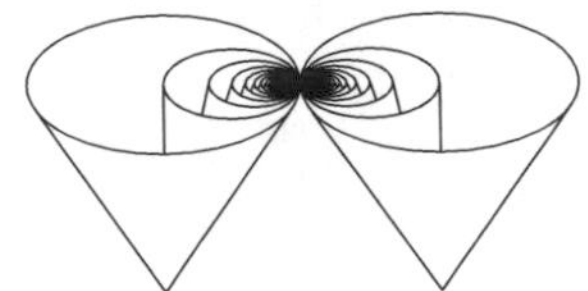

The space $\mathcal{W}$ can be equipped with a length metric (for example, the induced length metric from the shown embedding).

Show that $\mathcal{V}$ is simply-connected, but $\mathcal{W}$ is not; use the van Kampen theorem.

If we delete from the earrings all small circles and repeat the construction, then the obtained double cone becomes simply-connected and remains close to $\mathcal{W}$. That is, $\mathcal{W}$ is a Gromov–Hausdorff limit of simply-connected spaces.

Remark Note that the limit space in ((b)) does not admit a nontrivial covering.

5.16; (a) Suppose that a metric on $\mathbb{S}^2$ is close to the unit disk $\mathbb{D}^2$. Show that $\mathbb{S}^2$ contains a circle γ that is close to the boundary curve of $\mathbb{D}^2$. By the Jordan curve theorem, γ cuts $\mathbb{S}^2$ into two disks, say D_1 and D_2.

By 5.15(a), the Gromov–Hausdorff limits of D_1 and D_2 have to contain the whole $\mathbb{D}^2$; otherwise, the limit would admit a nontrivial covering.

Consider points $p_1 \in D_1$ and $p_2 \in D_2$ that are close to the center of $\mathbb{D}^2$. On one hand, the distance $|p_1 - p_2|_{\mathbb{S}^2}$ has to be small. On the other hand, any curve from p_1 to p_2 must cross γ, so its length is about 2 (or larger)—a contradiction.

(b) Show that one can remove fine tunnels from the standard 3-ball in such a way that: (1) the topology does not change, (2) the induced length metric is very close to the original one, and (3) the tunnels come sufficiently close to any point in the ball.

Consider the doubling of the obtained ball along its boundary; that is, two copies of the ball with glued corresponding points on their boundaries. The obtained space is homeomorphic to $\mathbb{S}^3$. Observe that the obtained space is sufficiently close to the original ball.

Source Dmitri Burago, Yuri Burago and Sergei Ivanov [13, Exercises 7.5.13 and 7.5.17].

5.18. Apply 1.14.
5.19. Let μ be a C-doubling measure on a space $\mathcal{X}$ from $\boldsymbol{Q}(C, D)$. Without loss of generality, we may assume that $\mu(\mathcal{X}) = 1$.

The doubling condition implies that

$$\mu[\mathrm{B}(p, \tfrac{D}{2^n})] \geqslant \tfrac{1}{C^n}$$

for any point $x \in \mathcal{X}$. It follows that

$$\operatorname{pack}_{\frac{D}{2^n}} \mathcal{X} \leqslant C^n.$$

By 1.14, for any $\varepsilon \geqslant \frac{D}{2^{n-1}}$, the space $\mathcal{X}$ admits an ε-net with at most C^n points. Whence $\boldsymbol{Q}(C, D)$ is uniformly totally bounded.
5.20. Since $\operatorname{diam} \overline{\mathrm{B}}[x, r] \leqslant r$, diameter of any space in $\boldsymbol{B}_{\mathcal{X}}$ is at most 2.

Suppose $\mathcal{X}$ is not doubling. Show that $n(x, r) = \operatorname{pack}_{\frac{r}{2}} \overline{\mathrm{B}}[x, r]$ is unbounded; that is, $n(x_n, r_n) \to \infty$ for some sequences x_n and $r_n > 0$. Conclude that $\boldsymbol{B}_{\mathcal{X}}$ is not uniformly totally bounded.

Suppose $\mathcal{X}$ is M-doubling. Show that

$$\operatorname{pack}_{\frac{r}{2^n}} \overline{\mathrm{B}}[x, r] \leqslant M^n,$$

and apply 5.18.
5.21; (a) Choose $\varepsilon > 0$. Since $\mathcal{Y}$ is compact, we can choose a finite ε-net $\{y_1, \ldots, y_n\}$ in $\mathcal{Y}$.

Suppose $f : \mathcal{X} \to \mathcal{Y}$ be a distance-noncontracting map. Choose one point x_i in each nonempty subset $B_i = f^{-1}[\mathrm{B}(y_i, \varepsilon)]$. Note that the subset B_i has diameter at most $2 \cdot \varepsilon$ and

$$\mathcal{X} = \bigcup_i B_i.$$

Therefore, the set of points $\{x_i\}$ is a $2 \cdot \varepsilon$-net in $\mathcal{X}$.
(b) Let $\boldsymbol{Q}$ be a uniformly totally bounded family of spaces. Suppose that each space in $\boldsymbol{Q}$ has an $\frac{1}{2^n}$-net with at most M_n points; we may assume that $M_0 = 1$.

Consider the space $\mathcal{Y}$ of all infinite integer sequences $m_0, m_1, \ldots$ such that $1 \leqslant m_n \leqslant M_n$ for any n. Given two sequences $\boldsymbol{\ell} = (\ell_1, \ell_2, \ldots)$, and $\boldsymbol{m} = (m_1, m_2, \ldots)$ of points in $\mathcal{Y}$, set

$$|\boldsymbol{\ell} - \boldsymbol{m}|_{\mathcal{Y}} = \tfrac{C}{2^n},$$

where n is the minimal index such that $\ell_n \neq m_n$ and C is a positive constant.

Observe that $\mathcal{Y}$ is compact. Indeed, it is complete, and the sequences with constant tails, starting from index n, form a finite $\tfrac{C}{2^n}$-net in $\mathcal{Y}$.

Given a space $\mathcal{X}$ in $\boldsymbol{Q}$, choose a sequence of $\tfrac{1}{2^n}$-nets $N_n \subset \mathcal{X}$ for each n. We can assume that $|N_n| \leqslant M_n$; let us label the points in N_n by $\{1, \dots, M_n\}$. Consider the map $f \colon \mathcal{X} \to \mathcal{Y}$ defined by $f : x \mapsto (m_1(x), m_2(x), \dots)$, where $m_n(x)$ is the label of a point in N_n that lies at the distance $< \tfrac{1}{2^n}$ from x.

If $\tfrac{1}{2^{n-2}} \geqslant |x - x'|_{\mathcal{X}} > \tfrac{1}{2^{n-1}}$, then $m_n(x) \neq m_n(x')$. It follows that $|f(x) - f(x')|_{\mathcal{Y}} \geqslant \tfrac{C}{2^n}$. In particular, if $C > 10$, then

$$|f(x) - f(x')|_{\mathcal{Y}} \geqslant |x - x'|_{\mathcal{X}}$$

for any $x, x' \in \mathcal{X}$. That is, f is a distance-noncontracting map $\mathcal{X} \to \mathcal{Y}$.

5.24. Let $\mathcal{K}$ be a compact space. Denote by $a(\mathcal{K})$ the largest diameter of connected component in a compact space $\mathcal{K}$. Further, let

$$b(\mathcal{K}) = \max_{p \in \mathcal{K}} \min_{q \neq p} \{|p - q|_{\mathcal{K}}\}.$$

Note that $b(\mathcal{K}) = 0$ if and only if $\mathcal{K}$ has no isolated points.

Show that if $a(\mathcal{K}) = b(\mathcal{K}) = 0$, then $\mathcal{K}$ is homeomorphic to the Cantor set.

Further, show that the sets

$$A_\varepsilon = \{\, \mathcal{K} \in \mathrm{GH} \; : \; a(\mathcal{K}) < \varepsilon \,\}$$

$$B_\varepsilon = \{\, \mathcal{K} \in \mathrm{GH} \; : \; b(\mathcal{K}) < \varepsilon \,\}$$

are open and dense in GH. Apply 1.10.

5.25; (a) Show that (the isometry classes of) finite metric spaces with only rational distances form a countable dense subset in GH.

(b)+(c) Choose two compact metric spaces $\mathcal{X}$ and $\mathcal{Y}$. Let $\mathcal{W} \supset \mathcal{X}', \mathcal{Y}'$ be as in 5.11; so $\mathcal{X}' \overset{iso}{=\!=} \mathcal{X}$, $\mathcal{Y}' \overset{iso}{=\!=} \mathcal{Y}$, and

$$\ell = |\mathcal{X}' - \mathcal{Y}'|_{\mathrm{Haus}\,\mathcal{W}} = |\mathcal{X} - \mathcal{Y}|_{\mathrm{GH}}$$

for some $\ell \geqslant 0$.

We can assume that $\mathcal{W} = \mathcal{X}' \cup \mathcal{Y}'$, so $\mathcal{W}$ is compact. Choose a compact geodesic extension $\mathcal{G}$ of $\mathcal{W}$; it exists by 2.2 (or 3.20).

Given $t \in [0, \ell]$, consider the set

$$\mathcal{Z}_t = \left\{ w \in \mathcal{G} \; : \; \mathrm{dist}_{\mathcal{X}'} w \leqslant t, \; \mathrm{dist}_{\mathcal{Y}'} w \leqslant \ell - t \right\}.$$

Observe that $t \mapsto \mathcal{Z}_t$ is a geodesic in Haus($\mathcal{G}$) from $\mathcal{X}'$ to $\mathcal{Y}'$. Conclude that $t \mapsto [\mathcal{Z}_t]$ is a geodesic in GH from $[\mathcal{X}]$ to $[\mathcal{Y}]$.

Source Alexander Ivanov and Alexey Tuzhilin [30].

5.26; (a) To check that $|{*} - {*}|_{GH'}$ is a metric, it is sufficient to show that

$$|\mathcal{X} - \mathcal{Y}|_{GH'} = 0 \quad \Longrightarrow \quad \mathcal{X} \overset{iso}{=\!=} \mathcal{Y};$$

the remaining conditions are trivial.

If $|\mathcal{X} - \mathcal{Y}|_{GH'} = 0$, then there is a sequence of maps $f_n : \mathcal{X} \to \mathcal{Y}$ such that

$$|f_n(x) - f_n(x')|_{\mathcal{Y}} \geqslant |x - x'|_{\mathcal{X}} - \tfrac{1}{n}.$$

Choose a countable dense subset $S \subset \mathcal{X}$ and pass to a subsequence such that $f_n(x)$ converges for any $x \in S$; denote by $f_\infty : S \to \mathcal{Y}$ the limit map. Note that f_∞ is distance-noncontracting, and it can be extended to a distance-noncontracting map $f_\infty : \mathcal{X} \to \mathcal{Y}$.

The same way we can construct a distance-noncontracting map $g_\infty : \mathcal{Y} \to \mathcal{X}$.

By 1.15, the compositions $f_\infty \circ g_\infty : \mathcal{Y} \to \mathcal{Y}$ and $g_\infty \circ f_\infty : \mathcal{X} \to \mathcal{X}$ are isometries. Therefore, f_∞ and g_∞ are isometries as well.

(b) The implication

$$|\mathcal{X}_n - \mathcal{X}_\infty|_{GH} \to 0 \quad \Rightarrow \quad |\mathcal{X}_n - \mathcal{X}_\infty|_{GH'} \to 0$$

follows from 5.10(a).

Now suppose $|\mathcal{X}_n - \mathcal{X}_\infty|_{GH'} \to 0$. Show that $\{\mathcal{X}_n\}$ is a uniformly totally bonded family.

If $|\mathcal{X}_n - \mathcal{X}_\infty|_{GH} \nrightarrow 0$, then we can pass to a subsequence such that $|\mathcal{X}_n - \mathcal{X}_\infty|_{GH} \geqslant \varepsilon$ for some $\varepsilon > 0$. By the Gromov selection theorem, we can assume that $\mathcal{X}_n$ converges in the sense of Gromov–Hausdorff. From the first implication, the limit $\mathcal{X}'_\infty$ has to be isometric to $\mathcal{X}_\infty$; on the other hand, $|\mathcal{X}'_\infty - \mathcal{X}_\infty|_{GH} \geqslant \varepsilon$—a contradiction.

5.28. Apply 2.20 and 5.27.

6.2. Let $F = \{\, n \in \mathbb{N} \;:\; f(n) = n \,\}$; we need to show that $\omega(F) = 1$.

Consider an oriented graph Γ with vertex set $\mathbb{N} \setminus F$ such that m is connected to n if $f(m) = n$. Show that each connected component of Γ has at most one cycle. Use it to subdivide vertices of Γ into three sets S_1, S_2, and S_3 such that $f(S_i) \cap S_i = \varnothing$ for each i.

Conclude that $\omega(S_1) = \omega(S_2) = \omega(S_3) = 0$, and hence,

$$\omega(F) = \omega(\mathbb{N} \setminus (S_1 \cup S_2 \cup S_3)) = 1.$$

Source The presented proof was given by Robert Solovay [60], but the key statement is due to Miroslav Katětov [34].

6.6. Choose a nonprincipal ultrafilter ω and set $L(\boldsymbol{s}) = s_\omega$. It remains to observe that L is linear.

Remark This construction identifies ultrafilters with vectors in $(\ell^\infty)^*$. Recall that $\ell^\infty = (\ell^1)^*$ and $\ell^1 \subsetneq (\ell^\infty)^*$. A principle ultrafilter is a basis vector in ℓ^1; nonprincipal ultrafilters lie in $(\ell^\infty)^* \setminus \ell^1$. The set of ultrafilters is the closure of basis vectors in ℓ^1 with respect to weak*-topology on $(\ell^\infty)^*$.

6.7. Apply 6.2.

6.11. Let γ be a path from p to q in a metric tree $\mathcal{T}$. Assume that γ contains a point x on distance ℓ from $[pq]$. Then

$$\text{length}\,\gamma \geqslant |p-q| + 2\cdot\ell. \tag{①}$$

Suppose that $\mathcal{T}_n$ is a sequence of metric trees that ω-converges to $\mathcal{T}_\omega$. By 6.10, the space $\mathcal{T}_\omega$ is geodesic.

The uniqueness of geodesics follows from ①. Indeed, if for a geodesic $[p_\omega q_\omega]$, there is another geodesic γ_ω connecting its ends, then it has to contain a point $x_\omega \notin [p_\omega q_\omega]$. Choose sequences $p_n, q_n, x_n \in \mathcal{T}_n$ such that $p_n \to p_\omega$, $q_n \to q_\omega$, and $x_n \to x_\omega$ as $n \to \omega$. Then

$$\begin{aligned}
|p_\omega - q_\omega| &= \text{length}\,\gamma \geqslant \\
&\geqslant \lim_{n\to\omega}(|p_n - x_n| + |q_n - x_n|) \geqslant \\
&\geqslant \lim_{n\to\omega}(|p_n - q_n| + 2\cdot\ell_n) = \\
&= |p_\omega - q_\omega| + 2\cdot\ell_\omega.
\end{aligned}$$

Since $x_\omega \notin [p_\omega q_\omega]$, we have that $\ell_\omega > 0$—a contradiction.

It remains to show that any geodesic triangle $\mathcal{T}_\omega$ is a tripod. Consider the sequence of centers of tripods m_n for given sequences of points $x_n, y_n, z_n \in \mathcal{T}_n$. Observe that its ultralimit m_ω is the center of a tripod with ends at $x_\omega, y_\omega, z_\omega \in \mathcal{T}_\omega$.

6.12. Construct $\boldsymbol{X}$ and distance-preserving embeddings $\mathcal{X}_n \hookrightarrow \boldsymbol{X}$ that satisfy 5.29. Given $x_\infty \in \mathcal{X}_\infty$, choose a sequence $x_n \in \mathcal{X}_n$ such that $x_n \to x_\infty$ in $\boldsymbol{X}$. Let x_ω be the ω-limit of the sequence x_n in $\boldsymbol{X}$. Note that $x_\omega \in \mathcal{X}_\infty$. Show that the map $x_\infty \mapsto x_\omega$ is defined; that is, it does not depend on the choice of the sequence x_n. Further, show that the map $x_\infty \mapsto x_\omega$ is an isometry of $\mathcal{X}_\infty$. Make a conclusion.

6.13. Further, we consider $\mathcal{X}$ as a subset of $\mathcal{X}^\omega$.

(a) Follows directly from the definitions.

(b) Suppose $\mathcal{X}$ compact. Given a sequence $x_1, x_2, \ldots \in \mathcal{X}$, denote its ω-limit in $\mathcal{X}^\omega$ by x^ω and its ω-limit in $\mathcal{X}$ by x_ω.

Observe that $x^\omega = \iota(x_\omega)$. Therefore, ι is onto.

If $\mathcal{X}$ is not compact, we can choose a sequence x_n such that $|x_m - x_n| > \varepsilon$ for fixed $\varepsilon > 0$ and all $m \neq n$. Observe that

$$\lim_{n\to\omega} |x_n - y|_{\mathcal{X}} \geqslant \tfrac{\varepsilon}{2}$$

for any $y \in \mathcal{X}$. It follows that x_ω lies at the distance $\geqslant \frac{\varepsilon}{2}$ from $\mathcal{X}$.
(c) A sequence of points x_n in $\mathcal{X}$ will be called ω-bounded if there is a real constant C such that

$$|p - x_n|_{\mathcal{X}} \leqslant C$$

for ω-almost all n.

The same argument as in (b) shows that any ω-bounded sequence has its ω-limit in $\mathcal{X}$. Further, if (x_n) is not ω-bounded, then

$$\lim_{n\to\omega} |p - x_n|_{\mathcal{X}} = \infty;$$

that is, x_ω does not lie in the metric component of p in $\mathcal{X}^\omega$.
6.14. Let us show that cardinality of $\mathcal{X}^\omega$ is at least continuum—it is sufficient to construct a continuum family $\mathcal{A}$ sequences of points on $\mathcal{X}$ such that for any two sequences (a_n) and (b_n) in $\mathcal{A}$ the equality $a_n = b_n$ holds only for finitely many n.

To do this, let us identify points in $\mathcal{X}$ with nonnegative integers. Consider the set $\mathcal{A}$ of all sequences a_n such that $a_0 = 0$ and $a_{n+1} = a_n + \varepsilon_n \cdot 2^n$ where $\varepsilon_n \in \{0, 1\}$ for any n. Observe that $\mathcal{A}$ has cardinality continuum and distinct sequences in $\mathcal{A}$ have distinct ω-limits.

Show and use that the spaces $\mathcal{X}^\omega$ and $(\mathcal{X}^\omega)^\omega$ have discrete metrics and both have cardinality at most continuum.

A More Conceptual Construction of $\mathcal{A}$ Choose a compact metric space $\mathcal{K}$ with continuum points, say $\mathcal{K} = [0, 1]$. Identify $\mathcal{X}$ with a dense subset of $\mathcal{K}$. For any point $k \in \mathcal{K}$, choose a sequence $a_n \in \mathcal{X}$ that converges to k. Observe that the family of all these sequences meet the required condition.

6.15. Choose a bijection $\iota\colon \mathbb{N} \to \mathbb{N} \times \mathbb{N}$. Given a set $S \subset \mathbb{N}$, consider the sequence $S_1, S_2, \ldots$ of subsets in $\mathbb{N}$ defined by $m \in S_n$ if $(m, n) = \iota(k)$ for some $k \in S$. Set $\omega_1(S) = 1$ if and only if $\omega(S_n) = 1$ for ω-almost all n. It remains to check that ω_1 meets the conditions of the exercise.

Comment It turns out that $\omega_1 \neq \omega$ for any ι; see the post of Andreas Blass [10].
6.17. Arguing as in 6.16, we get a pair of points x and y in $\mathcal{X}$ such that

$$|p - x| + |x - y| + |y - q| = |p - q|,$$

and there is no midpoint between x and y in $\mathcal{X}$ (possibly $p = x$ and $q = y$). Note that it is sufficient to show that there is a continuum of distinct midpoints in $\mathcal{X}^\omega$ between x and y in $\mathcal{X}$.

Since $\mathcal{X}$ is a length space, we can choose a $\frac{1}{n}$-midpoint $m_n \in \mathcal{X}$ between x and y. Note that the sequence m_n contains no converging subsequence. Conclude that we may pass to a subsequence of m_n such that $|m_i - m_j| > \varepsilon$ for a fixed $\varepsilon > 0$ and any $i \neq j$.

Argue as in 6.14 to show that there is a continuum of distinct ω-limits of subsequences of m_n; each such limit is a midpoint between x and y.

6.18. Consider the infinite metric $\mathcal{T}$ tree with unit edges shown on the diagram. Observe that $\mathcal{T}$ is proper.

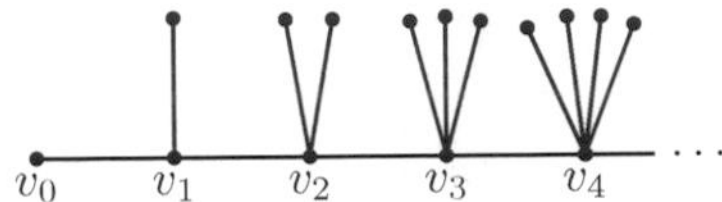

Consider the vertex $v_\omega = \lim_{n\to\omega} v_n$ in the ultrapower $\mathcal{T}^\omega$. Observe that ω has an infinite degree. Conclude that $\mathcal{T}^\omega$ is not locally compact.

6.19. Consider a product of an infinite sequence of two-point spaces.

Remark There are such examples of geodesic spaces with a cocompact isometric action of a finitely generated group [61].

6.20. Assume $\mathcal{L}$ is the Lobachevsky plane.

(a) Show that there is $\delta > 0$ such that sides of any geodesic triangle in $\mathcal{L}$ intersect a disk of radius δ. Conclude that any geodesic triangle in Asym$\mathcal{L}$ is a tripod.

(b) Observe that $\mathcal{L}$ is one-point-homogeneous and use it.

(c) By (b), it is sufficient to show that p_ω has a continuum degree.

Choose distinct geodesics $\gamma_1, \gamma_2 \colon [0, \infty) \to L$ that start at a point p. Show that the limits of γ_1 and γ_2 run in the different connected components of $(\mathrm{Asym}\mathcal{L})\backslash\{p_\omega\}$. Since there is a continuum of distinct geodesics starting at p, we get that the degree of p_ω is at least continuum.

On the other hand, the set of sequences of points in $\mathcal{L}$ has cardinality continuum. In particular, the set of points in Asym$\mathcal{L}$ has cardinality at most continuum. It follows that the degree of any vertex is at most continuum.

The proof for the Lobachevsky space goes along the same lines.

For the infinite three-regular tree, part (a) follows from 6.11. The three-regular tree is only vertex-homogeneous; the latter is sufficient to prove (b). No changes are needed in (c).

Remark The properties (b) and (c) describe the tree $\mathcal{T}$ up to isometry [17]. In particular, the asymptotic space of the Lobachevsky plane does not depend on the choice of the ultrafilter and the sequence $\lambda_n \to \infty$.

6.21. Denote by o_ω the point in $\mathrm{T}_o^\omega\mathcal{X}$ that corresponds to o. Argue as in (c) to show that $\mathrm{T}_o^\omega\mathcal{X} \setminus \{o_\omega\}$ has a continuum of connected components. Further, show that each connected component $\mathcal{W}_\alpha$ is isometric to $\mathbb{R} \times (0, \infty)$ with the metric described by

$$\begin{aligned}|(x_1, t_1) - (x_2, t_2)| &= \\ &= \min\{\, |(x_1, t_1) - (x_2, t_2)|_{\mathbb{R}^2}, t_1 + t_2 \,\}.\end{aligned}$$

Conclude that the space $T_o^\omega \mathcal{X}$ can be described as follows. Prepare continuum copies $\mathcal{W}_\alpha$ as above; denote by $(x,t)_\alpha$ the point in $\mathcal{W}_\alpha$ with coordinates (x,t). The tangent space is the disjoint union of single point o_ω and all $\mathcal{W}_\alpha$ with metric such that $|(x_1,t_1)_\alpha - (x_2,t_2)_\alpha|$ is the same as in $\mathcal{W}_\alpha$, and for the remaining pairs, we have $|o_\omega - (x,t)_\alpha| = t$ and $|(x_1,t_1)_\alpha - (x_2,t_2)_\beta| = t_1 + t_2$ if $\alpha \neq \beta$.

Bibliography

1. S. Alexander, V. Kapovitch, and A. Petrunin. *An invitation to Alexandrov geometry: CAT(0) spaces*. 2019.
2. S. Alexander, V. Kapovitch, and A. Petrunin. *Alexandrov geometry: foundations*. 2022. arXiv: 1903.08539 [math.DG].
3. N. Aronszajn and P. Panitchpakdi. "Extension of uniformly continuous transformations and hyperconvex metric spaces". *Pacific J. Math.* 6 (1956), 405–439.
4. P. Assouad. "Plongements lipschitziens dans $\mathbf{R}^n$". *Bull. Soc. Math. France* 111.4 (1983), 429–448.
5. V. N. Berestovskii and Y. G. Nikonorov. *On m-point homogeneous polytopes in Euclidean spaces*. 2022. arXiv: 2206.13096 [math.MG].
6. E. Bilokopytov. *Is it possible to connect every compact set?* MathOverflow. eprint: https://mathoverflow.net/q/359390.
7. R. H. Bing. "A homogeneous indecomposable plane continuum". *Duke Math. J.* 15 (1948), 729–742.
8. G. Birkhoff. "Metric foundations of geometry. I". *Trans. Amer. Math. Soc.* 55 (1944), 465–492.
9. W. Blaschke. *Kreis und Kugel*. 1916. Русский перевод: В. Бляшке, *Круг и шар*(1967).
10. A. Blass. *Selective ultrafilter and bijective mapping*. MathOverflow. eprint: https://mathoverflow.net/q/324261.
11. M. Bridson and A. Haefliger. *Metric spaces of non-positive curvature*. Vol. 319. Grundlehren der Mathematischen Wissenschaften. 1999.
12. T. Buehler. *Does there exist an isometry between L^p and ℓ^p?* MathOverflow. eprint: https://mathoverflow.net/q/112776.
13. D. Burago, Yu. Burago, and S. Ivanov. *A course in metric geometry*. Vol. 33. Graduate Studies in Mathematics. 2001.
14. H. Busemann. *Metric methods in Finsler spaces and in the foundations of geometry*. Annals of Mathematics Studies, No. 8. 1942.
15. P. Cameron. "The random graph". *The mathematics of Paul Erdős, II*. Vol. 14. Algorithms Combin. 1997, 333–351.
16. M. Deza and M. Laurent. *Geometry of cuts and metrics*. Vol. 15. Algorithms and Combinatorics. 1997. Русский перевод: М. Деза, М. Лоран, *Геометрия разрезов и метрик* (2001).
17. A. Dyubina and I. Polterovich. "Explicit constructions of universal $\mathbb{R}$-trees and asymptotic geometry of hyperbolic spaces". *Bull. London Math. Soc.* 33.6 (2001),727–734.
18. H. Federer. *Geometric measure theory*. Die Grundlehren der mathematischen Wissenschaften, Band 153. 1969. Русский перевод: Г. Федерер, *Геометрическая теория меры* (1987).

A. Petrunin, *Pure Metric Geometry*, SpringerBriefs in Mathematics,
https://doi.org/10.1007/978-3-031-39162-0

19. M. Fréchet. “Sur quelques points du calcul fonctionnel”. *Rendiconti del Circolo Matematico di Palermo (1884–1940)* 22.1 (1906), 1–72.
20. Z. Frolík. “Concerning topological convergence of sets”. *Czechoslovak Math. J* 10(85) (1960), 168–180.
21. M. Gromov. “Hyperbolic manifolds, groups and actions”. *Riemann Surfaces and Related Topics: Proceedings of the 1978 Stony Brook Conference*. Vol. 97. Ann. of Math. Stud. 1981, 183–213.
22. M. Gromov. *Metric structures for Riemannian and non-Riemannian spaces*. Modern Birkhäuser Classics. 2007.
23. O. Gross. “The rendezvous value of metric space”. *Advances in Game Theory*. Princeton Univ. Press, Princeton, N.J., 1964, 49–53.
24. F. Hausdorff. *Grundzüge der Mengenlehre*. 1914. Русский перевод: Ф. Хаусдорф *Теория множеств* (1937); English translation F. Hausdorff *Set Theory* (1957).
25. J. Heinonen. *Lectures on Lipschitz analysis*. Vol. 100. Report. University of Jyväskylä Department of Mathematics and Statistics. University of Jyväskylä, Jyväskylä, 2005, ii + 77.
26. D. Hilbert. „Ueber die gerade Linie als kürzeste Verbindung zweier Punkte.“*Math. Ann.* 46 (1895), 91–96.
27. T. Hu and W. A. Kirk. “Local contractions in metric spaces”. *Proc. Amer. Math. Soc.* 68.1 (1978), 121–124.
28. J. R. Isbell. “Six theorems about injective metric spaces”. *Comment. Math. Helv.* 39 (1964), 65–76.
29. J. R. Isbell. “Injective envelopes of Banach spaces are rigidly attached”. *Bull. Amer. Math. Soc.* 70 (1964), 727–729.
30. A. O. Ivanov, N. K. Nikolaeva, and A. A. Tuzhilin. “The Gromov–Hausdorff metric on the space of compact metric spaces is strictly intrinsic”. *Mat. Zametki* 100.6 (2016), 947–950.
31. A. O. Ivanov and A. A. Tuzhilin. “Isometry group of Gromov–Hausdorff space”. *Mat. Vesnik* 71.1-2 (2019), 123–154.
32. A. Karlsson. *Ergodic theorems for noncommuting random products*. url: http://www.unige.ch/math/folks/karlsson/.
33. A. Karlsson. *A metric fixed point theorem and some of its applications*. 2023. arXiv: 2207.00963 [math.FA].
34. M. Katětov. “On universal metric spaces”. *General topology and its relations to modern analysis and algebra, VI (Prague, 1986)*. Vol. 16. Res. Exp. Math. Heldermann, Berlin, 1988, 323–330.
35. B. Kleiner and B. Leeb. “Rigidity of quasi-isometries for symmetric spaces and Euclidean buildings”. *Inst. Hautes Études Sci. Publ. Math.* 86 (1997), 115–197 (1998).
36. B. Kirchheim, E. Spadaro, and L. Székelyhidi. “Equidimensional isometric maps”. *Comment. Math. Helv.* 90.4 (2015), 761–798.
37. C. Kuratowski. “Quelques probl‘emes concernant les espaces métriques nonséparables”. *Fundamenta Mathematicae* 25.1 (1935), 534–545.
38. U. Lang. “Injective hulls of certain discrete metric spaces and groups”. *J. Topol. Anal.* 5.3 (2013), 297–331.
39. U. Lang, M. Pavón, and R. Züst. “Metric stability of trees and tight spans”. *Arch. Math. (Basel)* 101.1 (2013), 91–100.
40. N. Lebedeva and A. Petrunin. *All-set-homogeneous spaces*. 2022. arXiv: 2211.09671 [math.MG]. to appear in St. Petersburg Math. J.
41. G. Lowther. *On the global structure of the Gromov–Hausdorff metric space*. MathOverflow. eprint: https://mathoverflow.net/q/212608.
42. J. Melleray. “Some geometric and dynamical properties of the Urysohn space”. *Topology Appl.* 155.14 (2008), 1531–1560.
43. K. Menger. “Untersuchungen über allgemeine Metrik”. *Math. Ann.* 100.1 (1928), 75–163.
44. B. Miesch. “Gluing hyperconvex metric spaces”. *Anal. Geom. Metr. Spaces* 3.1 (2015), 102–110.

45. B. Miesch and M. Pavón. "Weakly externally hyperconvex subsets and hyperconvex gluings". *J. Topol. Anal.* 9.3 (2017), 379–407.
46. B. Miesch and M. Pavón. *Ball intersection properties in metric spaces*. 2016. arXiv: 1610.03307 [math.MG]. to appear in J. Topol. Anal.
47. A. Nabutovsky. *Linear bounds for constants in Gromov's systolic inequality and related results*. 2019. arXiv: 1909.12225 [math.MG].
48. F. Nazarov. *Intrinsic metric with no geodesics*. MathOverflow. eprint: http://mathoverflow.net/q/15720.
49. A. Petrunin. *Center of convex figure*. MathOverflow. eprint: https://mathoverflow.net/q/432694.
50. A. Petrunin. *PIGTIKAL (puzzles in geometry that I know and love)*. AMR Research Monographs, Volume 2. 2022.
51. A. Petrunin. *m-point-homogeneous, but not $(m+1)$-point-homogeneous*. MathOverflow. eprint: https://mathoverflow.net/q/431426.
52. A. Petrunin. *Short selection in the space of subsets*. MathOverflow. eprint: https://mathoverflow.net/q/435157.
53. A. Petrunin and S. Stadler. "Metric-minimizing surfaces revisited". *Geom. Topol.* 23.6 (2019), 3111–3139.
54. A. Petrunin and A. Yashinski. "Piecewise isometric mappings". *St. Petersburg Math. J.* 27.1 (2016), 155–175.
55. A. Petrunin and S. Zamora Barrera. *What is differential geometry: curves and surfaces*. 2020. arXiv: 2012.11814 [math.HO].
56. A. Petrunin. *Metric geometry on manifolds: two lectures*. 2020. arXiv: 2010. 10040 [math.DG].
57. А. В. Погорелов. *Четвертая проблема Гильберта*. 1974. English translation: A. Pogorelov, *Hilbert's fourth problem* (1979).
58. K. Przesławski and D. Yost. "Continuity properties of selectors and Michael's theorem". *Michigan Math. J.* 36.1 (1989), 113–134.
59. W. Rudin. "Homogeneity problems in the theory of Čech compactifications". *Duke Math. J.* 23 (1956), 409–419.
60. R. M. Solovay. *Maps preserving measures*. 2011. url: https://math.berkeley.edu/~solovay/Preprints/Rudin_Keisler.pdf.
61. S. Thomas and B. Velickovic. "Asymptotic cones of finitely generated groups". *Bull. London Math. Soc.* 32.2 (2000), 203–208.
62. J. Tits. "Sur certaines classes d'espaces homog'enes de groupes de Lie". *Acad. Roy. Belg. Cl. Sci. Mém. Coll. in 8°*29.3 (1955), 268.
63. P. Urysohn. "Sur un espace métrique universel". *Bull. Sci. Math* 51.2 (1927), 43–64. Русский перевод в П. С. Урысон *Труды по топологии другим областям математики*, Том II, (1951) 747–777.
64. V. Uspenskij. "The Urysohn universal metric space is homeomorphic to a Hilbert space". *Topology Appl.* 139.1–3 (2004), 145–149.
65. J. Väisälä. "A proof of the Mazur–Ulam theorem". *Amer. Math. Monthly* 110.7 (2003), 633–635.
66. A. M. Vershik. "Random metric spaces and universality". *Uspekhi Mat. Nauk* 59.2(356) (2004), 65–104.
67. R. A. Wijsman. "Convergence of sequences of convex sets, cones and functions. II". *Trans. Amer. Math. Soc.* 123 (1966), 32–45.
68. R. Züst. *The Riemannian hemisphere is almost calibrated in the injective hull of its boundary*. 2021. arXiv: 2104.04498 [math.DG].

Index

A. Petrunin, *Pure Metric Geometry*, SpringerBriefs in Mathematics,
https://doi.org/10.1007/978-3-031-39162-0